THE MYSTERY OF
2012

PREDICTIONS, PROPHECIES
& POSSIBILITIES

SOUNDS TRUE
Boulder, Colorado

Sounds True, Inc.
Boulder, CO 80306

Compilation copyright © 2007, 2009 by Sounds True

SOUNDS TRUE is a trademark of Sounds True, Inc.
All rights reserved. No part of this book may be used or reproduced
in any manner without written permission from the author and publisher.

Book design by Rachael Tomich
Illustrations by Pam Uhlenkamp
Illustrations in "2012, Galactic Alignment, and the Great Goddess" are
    re-creations of artwork by John Nichols

ISBN 978-1-59179-674-9

Printed in Canada

Library of Congress Cataloging-in-Publication Data
The Mystery of 2012: predictions, prophecies, and possibilities.
p. cm.
ISBN 978-1-59179-611-4 (hardcover)
1. Prophecies (Occultism) 2. Mayas—Prophecies. I. Sounds True (Firm)
II. Title: The Mystery of 2012
BF1791.A15 2007
133.3—dc22
Library of Congress Control Number 2007007344

10 9 8 7 6 5

✪ This book is printed on recycled paper containing 100% post-consumer
waste and processed without chlorine.

Contents

Part Two

SCIENCE, BUSINESS, AND POLITICS
IN THE CONTEXT OF 2012

Part Three

SPIRITUALITY, SIGNS, AND
SYMBOLISM SURROUNDING 2012

Part Four

A NEW HUMANITY: EVOLUTION
TOWARD 2012 AND BEYOND

Introduction

TAMI SIMON

I first heard about the mystery of 2012 in the mid-1980s when I interviewed Mayan calendar expert José Argüelles for Boulder County Public Radio. The theme of the twelve-part series was "Earth Shift," and its focus was the Harmonic Convergence of August 16–17, 1987. (A transcript excerpted from this series begins on page 67.) According to José, this convergence marked Planet Earth's initiation into a twenty-five-year countdown that would bring us, in the year 2012, to the end of the Mayan Calendar's Great Cycle. José described 2012 as a period of dramatic evolutionary change that would affect the entire planet, a time of what he called "galactic synchronization," when Planet Earth would go through an astronomical shift that would bring it into alignment with powerful forces of change.

A pragmatist by nature, I wasn't immediately convinced. Still, I had to admit a deep curiosity. I felt somehow enlivened by the possibilities he was describing, and I had a lot of questions. Did all experts on the Mayan calendar agree about the 2012 end-date and its significance? Had the Mayan calendar proven a reliable prophetic tool in the past? What was "dramatic evolutionary change," and what might it feel and look like in the year 2012? Was the Earth truly in relationship with unseen forces—higher forces that had some kind of stake in our future? And, perhaps most important, was it possible for a world so burdened by factionalism and greed to make a significant, positive shift in such a short period of time?

I did not get answers to all these questions, and admittedly I didn't look very hard. The work of moment-by-moment transformation seemed much more compelling to me than theorizing about what might or might not happen in the future, and I turned my attention back toward spiritual practice, work, and the immediacies of life.

But, over the years, questions about 2012 have continued to resurface in my awareness, and with increasing urgency. Each year I meet more and more highly intelligent and serious spiritual practitioners who have seen, through direct inner vision, that the planet is being offered an opportunity to evolve during this 2012 window. They believe that 2012 offers a "gateway of possibility," an opportunity for awakening that requires our immediate attention. Moreover, a growing number of visionary writers and thinkers from diverse backgrounds—futurists, scientists, artists, eco-philosophers, and social theorists—are writing about 2012 and the decade that follows as a time of unprecedented and accelerated planetary change. Finally, 2012 is no longer some distant time in the future; it is just around the bend. If there was ever a time to investigate the possibilities of 2012 and prepare ourselves for its opportunities, that time is clearly now.

So, however much I might want to fend off formulating theories about the unknown and unknowable future, the mystery of 2012 feels like a phenomenon that is knocking at the door and asking for attention and clear thinking. Responding to this call, Sounds True has gathered this collection of essays on the mystery of 2012 from a wide variety of perspectives. Our goal has been to create an "investigative reader"—a book designed to aid the open-minded person in exploring the mystery of 2012 efficiently, thoroughly, and from a number of different vantage points.

Personally, I continue to be skeptical. I have even found myself wondering if it is misguided to set our sights on the possibilities of 2012 as a time of great collective transformation. For if we mistakenly interpret the 2012 predictions as permission to live our lives as a spectator sport—waiting for upcoming catastrophes or holding out for a galactic beam from another star system to solve all of our planet's problems—then the promise and possibility of 2012 has turned into an excuse for nonaction. Interestingly, this is a point highlighted by many of the contributors to this collection. In this book, 2012 is most often described as a choice point, a time of intensified possibility and opportunity, rather than an apocalyptic time bomb destined to explode at midnight on December 21, 2012. What these visionaries are calling us to do as we investigate 2012 is to be intensely alert, responsive, and creatively engaged with the possibilities that may be unfolding around us.

That said, I encourage you to embark on the journey of this anthology with an aroused sense of curiosity. I have found that the big-picture view of many 2012 visionaries can have a mind-expanding effect. Their writings have helped shift my own view away from the human, historical dimension of change, toward a vaster view—one replete with energetic, vibratory, and invisible dimensions that can infuse our human efforts with previously untapped energy. Opening to the possibilities of 2012 requires us to open to mystery, to open to the powers of the cosmos that are transhuman. It is my belief that these powers can inspire and support our actions.

In his essay, author Daniel Pinchbeck describes 2012 as a time for "The Return of Quetzalcoatl," the mythical feathered serpent of Mesoamerica that symbolizes the meeting of snake and bird, or Earth and Heaven. And it is just this kind of meeting that we need. Only with our feet firmly rooted in the soil of our immediate situation can we reach out and align with wisdom and power from beyond. May this collection of essays inspire this meeting in you, as we live together in the mystery of 2012.

Tami Simon
Publisher, Sounds True

Choice Point 2012

OUR DATE WITH THE WINDOW OF EMERGENCE

GREGG BRADEN

As they hear 2012 predictions, many wonder if there is any scientific evidence. Gregg Braden explores the possibility that 2012 will bring a reversal of the Earth's magnetic poles and what such a reversal may mean for the largest population the world has ever seen. In the following essay, using the Maya's incredible accomplishments and predictions as a starting point, he probes what we know today, looking for scientific evidence that the planet's magnetism will reverse and that sunspots or sun storms signify a coming change. Braden draws from physical evidence, quantum theory, and historic trends to gauge the likelihood of massive destruction or the emergence of an empowering new reality in 2012. After weighing the evidence, the choice, he says, is ours.

Does the ancient Mayan calendar hold the secret to an epic event that will occur within our lifetimes? If so, does that event hold the key to our future, and perhaps even our very survival? A growing body of evidence suggests that the answer to these questions is, Yes. Now we must ask ourselves, Why?

Why did an advanced civilization suddenly appear more than 1,500 years ago with the most sophisticated galactic clock known until modern times, build a massive civilization focused on expansive galactic cycles, and then disappear? Why does their calendar, which coincides with the 5,000 years of recorded human history, end on such a precise date—a date within our own lifetime? What is the significance of the winter solstice on December 21, 2012?

To answer these questions, we must cross the traditional boundaries that have separated science, religion, spirituality, and history and marry these many sources of knowledge into a single new wisdom. When we do, something remarkable begins to happen.

THE MAYAN MYSTERY

The Mayan civilization is an anomaly in the traditional view of history and culture. Archaeological records show that the first Maya "suddenly" appeared over a millennium and a half ago, in the remote areas of what is today Mexico's Yucatán Peninsula, Guatemala, and parts of Honduras and Belize. What sets the Mayan civilization apart from other cultures during the same period is that the Maya appear to have arrived with an advanced technology already in place, rather than evolving their technology over a long period of time, as we would expect. Although there are many theories, no one has solved definitively what is called the "Mayan riddle."

In his exploration of this ancient mystery, author Charles Gallenkamp summarizes the irony of the Mayan presence: "No one has satisfactorily explained where or when Mayan civilization originated or how it evolved in an environment so hostile to human habitation." He elaborates on how little we actually know about our ancient ancestors, observing that whatever it was that led to "the sudden abandonment of their greatest cities during the ninth century A.D.—one of the most baffling archaeological mysteries ever uncovered—is still deeply shrouded in conjecture."[1]

While experts may not agree on precisely *why* such a powerful civilization seems to have disappeared, they can't argue over the marvel of what it left behind. And any discussion of Mayan accomplishments must acknowledge what is arguably the single most sophisticated artifact of all: their unsurpassed calculation of cosmic cycles and time. Before the twentieth century, the Mayan calendar appears to have been the most sophisticated method of tracking galactic time. Even today, modern Maya keep track of galactic time, as well as local time, using the system that experts such as Michael D. Coe tell us has "not slipped one day in over 25 centuries."[2] In addition to tracking familiar solar and lunar cycles, the 5,000-year-old Mayan calendar appears to track something even more surprising: the rare celestial alignment of our solar system, our sun, and our planet with the center of our galaxy—an event that will not happen again for another 26,000 years.

The key to the Maya's "galactic timer" was a 260-day count called the *tzolkin*, or "Sacred Calendar," which was intermeshed with another 365-day calendar called the "Vague Year." The Maya viewed these two cycles of time progressing like the cogs of two wheels, turning until the rare moment when one day on the Sacred Calendar matched the same day on the Vague Year. That rare and powerful day marked the end of a 52-year cycle and was part of an even greater expanse of time called the "Great Cycle." According to this explanation and the tradition of the Mayan priests themselves, records indicate that the last Great Cycle began in August of 3114 B.C.E.—the approximate time of the first Egyptian hieroglyphics—and ends in the near future, in the year 2012 C.E. Specifically, the cycle ends on December 21, 2012, when our sun will move into direct alignment with the equator of the Milky Way galaxy.

Scientists acknowledge that this galactic alignment will occur and that the Mayan calendar marks the event. The question most often asked is simply, What does it mean? There are those who discount the phenomenon as little more than an interesting oddity that we will be lucky to see during our lifetime. Others suggest that the close of the Great Cycle marks the convergence of rare cosmic processes, with implications that range from joyous to frightening.

Dr. José Argüelles, an expert in Mayan cosmology, suggests that the first years of the new millennium are part of a subcycle that began in 1992, marking the emergence of what he calls "nonmaterialistic, ecologically harmonic technologies . . . to compliment the new decentralized mediarchy information society."[3]

Using the same information, others warn that the end of the Mayan calendar coincides with celestial events that may hold profound, and even dangerous, consequences for life on Earth as we know it. An electronic magazine based in India, for example, carried an article in the March 1, 2005, edition, describing the results of the Hyderabad Computer Model for a polar shift to coincide with the calendar's end-date. The frightening headlines stated, "Computer models predict magnetic pole reversal in Earth and sun can bring end to human civilization in 2012." The article also described a worse-case scenario of what a world without a magnetic field could mean.[4]

While there are many ideas of what we may expect as the end-date of the Mayan calendar draws near, most people feel that *something* is going to happen—the question is, What? What were the Mayan timekeepers trying to tell us about a date that none of them would even live to see? Because 2012 is only a few short years away, and happens to coincide with unprecedented changes happening in our solar system, a growing number of scientists suggest that it is in our best interest to understand what the Mayan timekeepers were trying to tell us. Perhaps the best way to discover what we *don't* know is to take a look at what we *do* know.

EARTH'S MAGNETISM: A SURE THING?

When we think of things that are certain in life, we tend to count the magnetic fields of our planet among those certainties. For as long as anyone living today can remember, every time we have looked at the needle of a compass, the tip of the needle has pointed in the same direction—"up there," toward the magnetic north pole of the Earth. And while we tend to think of Earth's north and south poles as a sure thing, the reality is that our planet's magnetism is anything but certain.

We know, for example, that every once in a while something truly mind-boggling, almost unthinkable, happens. For reasons that are still not fully understood, our familiar north and south poles trade places—the magnetic field of the Earth does a complete flip-flop. Although polar reversals are rare in the history of civilizations, the geologic record shows that they happen routinely in terms of Earth's history. Magnetic reversals have already happened 171 times in the last 76 million years, with at least 14 of those reversals occurring in the last 4.5 million years alone.[5]

And while they are definitely cyclic, the reversals appear to vary in time, making the time of the next one an uncertainty. There are symptoms, however, that precede the flip-flops, such as abrupt changes in weather patterns and a rapid weakening of the planet's magnetic field—both of which are happening right now. It is the appearance of these symptoms today, and the fact that we are "overdue" for a polar shift, that has led a growing number of mainstream scientists to suggest that we are in the early stages of just such a reversal.

In July 2004, the *New York Times* took the possibility of Earth's magnetic reversal seriously enough to dedicate an entire portion of its Science section to describing just what a magnetic reversal is and the possible implications of one taking place. The article stated, "The collapse of the Earth's magnetic field, which both guards the planet and guides many of its creatures, appears to have started in earnest about 150 years ago."[6] There is little doubt, at least in the minds of some scientists, that the reversal has already begun.

Geologic measurements do, in fact, show that Earth is declining from a peak of magnetic intensity 2,000 years ago and that the values have steadily dropped to a point 38 percent lower now than then. Measurements taken since the mid-1800s further support the fact of the magnetic decline, showing that planetary magnetism has lost 7 percent of its strength in the last 100 years alone.[7] Even though the symptoms of a polar reversal are present, they may not be easily recognized for a number of reasons, including the assumption that the reversal of Earth's magnetic fields happens as a long, slow process occurring at a constant rate.

New interpretations of the evidence present a strong case that something else may be happening. It may be that the more the field weakens, the faster it weakens. And if this is the case, past reversals may have occurred much faster than we currently believe. In the remote northern latitudes of Siberia, for example, wooly mammoths, believed to have been caught in a polar reversal during the last ice age, have been found frozen in midstride, with their last meal still in their mouths—proof that the abrupt climate change accompanying such a shift can happen really fast!

In 1993, *Science News* published a study describing how "the task of finding an accurate reversal record seems to be all the more difficult because the magnetic field weakens considerably when it switches direction," just as we are witnessing today.[8] The question remains: Just how weak does the field become before a reversal? While the answer to this question is not certain, what we do know is that such monumental events don't just happen by themselves on Earth. They appear to be linked to events happening to our celestial neighbors, and possibly even the entire galaxy.

OUR SUN: THE QUIET BEFORE THE GREAT STORM

Since Galileo's first telescopes allowed astronomers to observe the heavens, European astronomers have known that our sun experiences regular cycles of intense magnetic storms—sunspots—followed by predictable periods of quiet. These cycles have been observed on a regular basis since 1610. Since the measurements began, 23 cycles of sunspots, averaging 11 years each, have occurred, with the last beginning in May of 1996. Precisely when the most recent cycle would end was a mystery—that is, until the spring of 2006, when NASA reported the event that astronomers had been waiting for. On March 10 of that year, the sunspots and solar flares suddenly stopped, and the sun became "quiet," signaling the end of the current sunspot cycle. The quiet, however, is misleading.

The end of one cycle is the indication that a new cycle—and new storms—is beginning. What makes the coming cycle so different is that the strength of the sunspots observed from 1986 to 1996 suggests that the next cycle will be one of the most intense ever recorded. "The next sunspot cycle will be 30–50 percent stronger than the previous one," says Mausumi Dikpati of the National Center for Atmospheric Research (NCAR) in Boulder, Colorado.

David Hathaway, of the National Space Science and Technology Center, agrees and suggests that the sunspots created in the previous cycle are expected to amplify themselves and "reappear as big sunspots" in the new cycle. If so, the solar magnetic storms will be second in intensity only to those of 1958, when the aurora borealis illuminated the night sky as far south as Mexico. At that time, however, we didn't have our current communications technology, such as satellites, that can be disrupted by such storms. While the predicted intensity of the new solar cycle would normally be of concern, it becomes even more significant when we consider the projected date for the most intense portion of the storm—the solar maximum. Based on the 1986–1996 cycle, NCAR's Dikpati places the projected date for the solar maximum at 2012, coinciding with the Mayan calculations of our sun's galactic alignment. Because this phenomenon has never happened with the population and technology that we have today, no one knows for certain what effects these solar storms may have on our future.

WHAT DOES IT MEAN?

We know that the sun is going through a magnetic shift, and the Earth appears to be in the early stages of a polar reversal as well. What do these celestial events have to do with the Mayan calendar? What does it all mean? These are very good questions, and they are important because the one thing we know for certain is that all life is strongly influenced by magnetic changes. There are countless studies in scientific literature that describe how many species of animals, from whales and dolphins to hummingbirds and wildebeests, rely on Earth's magnetic "superhighways" to navigate their way to feeding and mating grounds. While we may not use these "superhighways" quite the same way, it appears that humans are no exception.

In 1993, an international team studying *magnetoreception*, the ability of our brains to detect magnetic changes in the Earth, announced a discovery that makes 2012, the Mayan cycles, and Earth's magnetic fields even more significant than previously thought. The team published the remarkable findings that the human brain contains "millions of tiny magnetic particles."[9] These particles connect us, just as they do other animals, to Earth's magnetic field in a powerful, direct, and intimate way. And this connection carries powerful implications. If Earth's magnetic fields are changing in the 2012 time frame, then we too are affected.

We know, for example, that magnetic fields have a profound influence on our nervous systems, our immune systems, and our perceptions of space, time, dreams, and even reality itself. And while the strength of our planet's magnetic field may be measured as a general reading, it varies locally from place to place. Early in the twentieth century, magnetic, ribbonlike patterns were charted and published by scientists as a contour map of the world.[10] The map displays the strength of the magnetic lines overlaying the continents and shows the places on Earth where the people of the world experience the strongest—and the weakest—effects of the planet's magnetic fields. To understand why this is important for the 2012 cycle, we need look no further than consciousness itself.

While there is a lot that we don't know about consciousness, there is one thing that we are certain of: the stuff that consciousness is made of is energy, and that energy includes electricity and magnetism. While the electromagnetic nature of consciousness is still being explored, it does appear that Earth's magnetics play a key role in how we accept new ideas and change in our lives. If we think of magnetic fields as a form of energetic "glue," we can use this metaphor as a possible explanation for why some parts of the world, and our own country, are slower to accept change, while other places seem to jump at the opportunity to try something new.

Our "magnetic glue" model suggests that places with stronger magnetic fields (more glue) are more deeply entrenched in tradition, beliefs, and existing ideas. In places where the fields are weaker, just the opposite is true. In these places, people seem compelled to create change. Although areas of low magnetic intensity may be ripe for something new, how that change is expressed is up to those living there. With this in mind, a look at our global magnetic map may help us make sense of places that seem continuously locked in conflict, as well as understand why innovation and change seem to spark in one area before spreading to others. Immediately, one link is so clear that it almost jumps off of the page.

The places in our world with the lowest magnetic intensity, such as the zero contour line (zero magnetic gauss), which runs directly beneath the Suez Canal and into Israel, are precisely the places where we see the greatest opportunities for change. Sometimes the change comes as a smooth transition from one idea to the next. Sometimes it comes as a struggle. In our Middle East example, we see the struggle that can result from the attempt to preserve ancient tradition in a place that compels change. The low magnetic field does not mean that the change must be expressed as conflict; it simply provides the conditions for new ways of looking at things. Once again, it is up to those living in these zones of opportunity how they express the change that the low fields invite.

A similar zero magnetic contour line exists parallel to America's West Coast. Not surprisingly, this is a part of the world that is also known as a hotbed for change, although it is expressed in a different way than in the Middle East. Stretching from southern California to northern Washington, this zone of low magnetic strength is often seen as a realm of new thought and innovation—and with that innovation, we find leading ideas in science, technology, fashion, music, and art.

On the opposite end of the spectrum, there are zones of extremely high magnetism where, historically, change comes about slowly over time, and often as the result of a struggle. In central Russia, for example, contour values are some of the highest in the world, with readings at more than 150 gauss units. Russia's recent experience with the political reorganization that coincided with the end of the Cold War is a beautiful example of how areas with high magnetic fields tend to cling to tradition and embrace change slowly, over a long period of time. However, it also shows that once change begins in these areas, it carries a momentum that makes itself known in a way that cannot be missed.

Even without such evidence, we know intuitively that we are affected by planetary magnetic forces. Any law enforcement officer or health-care practitioner will attest to the intense, and sometimes bizarre, behavior that is seen during a full moon. When the magnetic strength changes suddenly, it changes the way we feel, and that change can be disorienting if we don't understand why it happens. To those who do understand, however, such moments can be a powerful gift—the opportunity to release the patterns of belief that have caused pain in their lives, hurt in their families, and disease in their bodies, and to embrace new and life-affirming beliefs. Artists and musicians know this and often anticipate full-moon cycles as periods of great creativity. These existing examples of the connection between consciousness and magnetism offer important insights into how future changes in magnetic strength may affect us, as some expect them to, during the 2012 cycle.

Is it possible that by altering the way we feel about change in our lives, we can actually transform the way we experience things, like the frightening predictions of 2012? While there are a number of new discoveries that show us how consciousness directly affects our world, they are generally variations of the nearly century-old experiment designed to determine how much our beliefs really affect our reality.

In 1909, Geoffrey Ingram Taylor, a British physicist, devised the famous Double Slit Experiment and began a revolution in the way we view ourselves in the universe. The bottom line of his experiment was that the mere presence of consciousness in a room—*people*—affected the way the quantum particles (the stuff our world is made of) behaved.

On February 26, 1998, Taylor's experiment was repeated by scientists at Israel's Weizmann Institute of Science. Not only did they confirm that our world is affected just by observation, but they also discovered that "the greater the amount of 'watching,' the greater the observer's influence on what actually takes place."[11] In other words, the more those present focused on the experiment, the greater the influence their focus had on the outcome. And herein lies the key to understanding what quantum physics and the Mayan calendar may really be saying to us about our power in the universe.

Our world, our lives, and our bodies are the result of the quantum language of our beliefs, judgments, love, and fear regarding those things. The same theories that describe our reality as being influenced by our beliefs suggest that if we want to change our reality, we must change the way we see ourselves. In 1957, Princeton University physicist Hugh Everett III carried these ideas one step further, creating the theory that describes *how* the focus of our awareness creates reality.

In a landmark paper that included his Many Worlds Theory, Everett described simple moments in time when it becomes possible to "jump" from one reality to another by creating a quantum bridge between the two already-existing possibilities.[12] He called these windows of opportunity "choice points" and described them as the times when conditions make it possible to begin one path of experience and then to change the outcome of that experience by changing the focus of our awareness—our beliefs.

From this perspective, the 2012 end-date appears to represent a great cosmic choice point—the time in the galactic machinery when all of the parameters fall into place to make it easier for us to choose a new way of seeing ourselves in the universe and a new way of being.

CHOICE POINT 2012: BRINGING IT ALL TOGETHER

Now we have the information, and the language, that may answer our question as to the significance of December 21, 2012. From the 8,000-year-old Hindu Vedas and the 5,000-year-old Hebrew calendar to the indigenous prophecies of Asia and the Americas, ancient traditions suggest that something really big will happen during our time in history. It is important to note that there is nothing in any of the prophecies that tells us unconditionally that the world itself will end on this date. What they do say is that the world as we have known it will enter a time of change on this date. How we respond to that change will define how we experience it and our lives in the next age of our existence.

What sets the Mayan prophecy apart from the general predictions of other cultures is that it has an expiration date that occurs in our lifetime. The last cycle of the intangible Mayan calendar corresponds to a series of tangible events, some of which are already happening today. Here is what we know for certain:

- The end of the Mayan Great Cycle marks a rare alignment of our planet, our solar system, and the center of our galaxy—one that will not occur again for another 26,000 years.
- On March 10, 2006, a cycle of solar storms ended and a new cycle began. It is predicted to peak in 2012, with an intensity 30–50 percent greater than previous cycles.
- Scientists agree that Earth's magnetic fields are weakening quickly, and some suspect that we are in the early stage of a polar reversal.
- Correlations between the magnetic fields of the Earth and human experience suggest that it is easier for us to accept change and adapt to new ideas in weaker fields of magnetism.
- Recent validation of quantum principles proves that the way we perceive our world—our beliefs about our experience—strongly influences our physical reality.

With these facts in mind, the December 21 solstice of 2012 appears to be a great cosmic window of opportunity. The opportunity is for us to benefit from our individual and collective experiences of the past 5,000 years of history, while releasing the destructive beliefs that have created disease in our bodies and limited our life span, led to the differences that have separated us, justified the great wars of history, and destroyed portions of our world. Such a possibility exists only when all of the parameters needed for such a change converge into a single window of time. The 2012 end-date appears to be just such a time.

If (1) the presence of low magnetic fields primes us to adapt to change and accept new experiences, (2) the loss of Earth's magnetic shield allows the direct influence of high-energy fields that are normally deflected to bring change, and (3) both are happening in the general time of the 26,000-year galactic alignment, then it is as if the cosmos has conspired to shower us with the power of what José Argüelles called the galactic "synchronization beam." What a rare and bizarre opportunity. How we receive this opportunity is the choice that we are making today.

In 2005, *Scientific American* published a special edition titled *Crossroads for Planet Earth,* which identified a number of scenarios that, left unchecked, hold the potential to end life on Earth as we know it today. The point of the issue was that while any one of the scenarios is catastrophic, they are all happening today. Clearly the choices of war, depletion of resources, genocide, technology abuse, and fossil fuel technology are not sustainable for another millennium. To change the way we live, we must change the way we believe. Such a change can only come from a powerful and holistic view of who we are in the universe. Of the many things that 2012 represents, it may offer an unprecedented window of precisely such an opportunity.

Are we heading toward a time of unprecedented catastrophe, 1,000 years of peace, or both? No one knows for sure. While the solar cycles and magnetic reversals are very real and have definitely happened in our ancient past, they have never happened while 6.5 billion people lived on the planet, and certainly not with the majority of those people dependent upon the technology of power grids, communications, computers, and global positioning satellites. Just as uncertainty accompanies any birth, we simply don't know what our birth into such a monumental experience may mean to life, technology, our emotions, and our bodies.

We do know, however, that ancient humans may have experienced similar, though less-intense cycles, some that may have even occurred as recently as 10,000 years ago. While biblical and oral traditions suggest that such a time is certainly not "business as usual," the fact that people lived to record those cycles tells us that such events are survivable.[13] The new discoveries linking the physics of belief with reality also tell us that how we feel about our experience has a direct effect upon *what* we actually experience. For our date with 2012, this suggests that if we live life focused upon all of the bad things that may happen, we will miss the joyous experiences that may actually keep those bad things from happening.

Today, when you ask the Mayan descendants what happened to their ancestors, they will tell the story of the generation of timekeepers who, one day, left their temples, observatories, and pyramids, walked into the jungles, and simply "vanished," returning to the place from where they had come. Regardless of what their story means to us, it is clear that whoever those original Maya were, they knew something in their time that we are just beginning to understand in ours. The key to their message is that their secret was more than the precise representation of time on a stone slab. The one piece of their wisdom that they could not inscribe into their hieroglyphic message is precisely the piece that gives meaning to the end of their cycle. The piece of their wisdom is us and whether we can embrace the fact that we are a family in this world, very possibly part of a greater family in the cosmos, and that the "gestation" that José Argüelles describes as our experience in history is complete.

From these perspectives, the solstice of December 21, 2012, becomes a powerful window for our collective emergence into our greatest potential. Such a moment is so rare that we have been preparing for it for more than 5,000 years, and it will be another 26,000 years before the same opportunity cycles around again. As the source of our light—our sun—moves into perfect alignment with the center of our galaxy, are we ready to receive the greatest gift of all—the gift of our true selves? The stage is set, the choice is ours, the cosmos is waiting. Do we have the wisdom to marry science, history, tradition, and belief into the miracle that awaits us on December 21, 2012?

A Singularity in Time

PETER RUSSELL

As talk of 2012 becomes more widespread, questions arise: How does evolution speed up? Claims are made that time is accelerating—how is this possible? Can our state of being and our planet really change as quickly as some 2012 theories suggest? In what follows, Peter Russell, acclaimed author of From Science to God, *speaks to these questions and explains what he foresees happening in the coming years and in 2012. He applies the theories of evolution and time-change to technology, the planet, and our lives, in the next decade and beyond.*

The pace of life is forever speeding up. Technological breakthroughs spread through society in years rather than centuries. Calculations that previously would have taken decades are now made in minutes. Communication that used to take months happens in seconds. In almost every area of life, change is occurring faster and faster.

Yet this acceleration is not confined to modern times. Although medieval architecture and agriculture, for instance, varied very little over the period of a century, changes did occur much faster than they had in prehistoric times, when tools remained unchanged for thousands of years.

Nor is this quickening confined to humanity; it is a pattern that stretches back to the dawn of life on Earth. The first simple life-forms evolved nearly four billion years ago. Multicellular life appeared a billion or so years ago. Vertebrates with central nervous systems, several hundred million years ago. Mammals appeared tens of millions of years ago. The first hominids stood on the planet a couple of million years ago; Homo sapiens, a few hundred thousand years ago. Language and tool use emerged tens of thousands of years ago. Civilization, the movement into towns and cities, started a few thousand years ago. The Industrial Revolution began three centuries ago. Finally, the Information Revolution is but a few decades old. Each successive step happens sooner than the last.

WHY DOES EVOLUTION ACCELERATE?

The reason for this acceleration is that each new development is standing on the shoulders, so to speak, of what has come before. A good example is the advent of sexual reproduction some 1.5 billion years ago. Until that time, cells reproduced by splitting into two, with each of the new "sisters" being exact clones of the original. With sexual reproduction, two cells came together, shared genetic information, and produced offspring containing a combination of their genes. It no longer took many generations for one genetic difference to arise. Differences now occurred in every generation, speeding evolution a thousandfold.

A more recent example is the transition from the Industrial Age to the Information Age. When it came to manufacturing computers, we did not need to reinvent factories or global distribution systems; that expertise had already been gained. We had simply to apply it to the production of computers. Thus, the Information Revolution established itself much more quickly.

This pattern is set to continue in the future—each new phase requiring a fraction of the time required in the previous phase. In the future, we might expect the same amount of change we've seen in the last 20 years to take place in years rather than decades.

It is difficult, therefore, to predict what the world will be like in 10 or 20 years. Two hundred years ago, no one predicted we would have telephones or movies, let alone cell phones or the Internet. Just 20 years ago, very few of us had any notion of the World Wide Web or of how dramatically it would change our lives. Similarly, who knows what new breakthroughs or developments will be transforming our lives 10 years from now?

APPROACHING A SINGULARITY

So where is all this leading? Some people think we are headed toward what is called a "singularity." This is the term that mathematicians give to a point when an equation breaks down and ceases to have any useful meaning. The rules change. Something completely different happens.

A simple example of a singularity occurs if you try to divide a number by zero. If you divide by smaller and smaller numbers, the results will be larger and larger numbers. But if you divide a number by zero, you get infinity, which is not a number in the everyday sense. The equation has broken down.

The idea that there might be a singularity in human development was first suggested by the mathematician Vernor Vinge and subsequently by others, most notably Ray Kurzweil in his book *The Singularity Is Near*. They argue that if computing power keeps doubling every 18 months, as it has done for the last 50 years, then sometime in the 2020s there will be computers that can equal the performance of the human brain. From there, it is only a small step to a computer that can surpass the human brain. There would then be little point in our designing

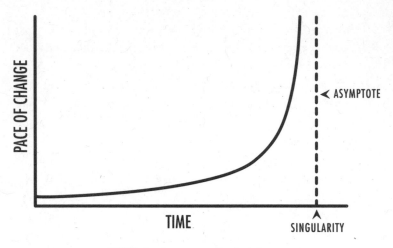

FIGURE 1: Growth curve approaching singularity.

future computers; ultraintelligent machines would be able to design better ones, and do so faster.

What happens then? is a big question. Some propose that humans would become obsolete; machines would become the vanguard of evolution. Others think there would be a merging of human and machine intelligence—downloading our minds into computers, perhaps. The only thing we can predict confidently is that this would be a complete break from the patterns of the past. Evolution would move into a radically new realm.

But this transition, as major as it would be, would not yet be a true singularity in the mathematical sense. Evolution— whether human, machine, or a synthesis of the two—would continue at an ever-increasing pace. Development timescales would continue to shorten, from decades to years, to months, to days. Before long, they would approach zero. The rate of change would then become infinite. We would reach a true mathematical singularity.

The idea that humanity is heading toward a point of infinitely rapid change was explored by Terence McKenna in his book *The Invisible Landscape*. He developed a mathematical fractal function, which he called the "timewave," that appeared to match the overall rate of ingression of novelty in the world. ("Ingression of novelty" is a term coined by the philosopher Alfred North Whitehead to denote new forms or developments coming into existence.) This timewave is not a smooth curve, but one that has peaks and troughs corresponding to the peaks and troughs of the rate of ingression of novelty across human history.

The most significant characteristic of McKenna's timewave is that its shape repeats itself, but over shorter and shorter intervals of time. The curve shows a surge in novelty around 500 B.C., when Lao Tzu, Plato, Zoroaster, Buddha, and others were exerting a major influence on the millennia to come. The repeating nature of McKenna's timewave shows the same pattern occurring in the late 1960s, where it happened 64 times faster. In 2010, the pattern repeats again, 64 times faster still, and in 2012, 64 times faster still. The timescale is compressed from months to weeks to days, tending very rapidly toward zero: a point McKenna called "Timewave Zero."

But when precisely is this date? McKenna experimented with sliding his curve up and down history to look for a best fit. Eventually, he chose December 22, 2012. At the time, he did not know that the Mayan calendar also ended its 5,124 years one day earlier. McKenna himself was not overly attached to the date; he confided that he would be intrigued, come 2012, to see whether his conjectures about infinite novelty would indeed prove correct. Sadly, he passed away in 2000.

Personally, I am not so concerned with what actually will or will not happen on that precise date of December 21, 2012.

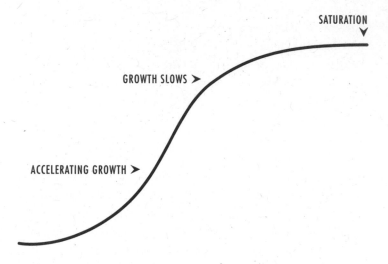

FIGURE 2: As any growth reaches saturation, its rate slows and levels out.

Indeed, almost every prediction ever made that related to a specific date has failed to materialize. I am more interested in where this accelerating pattern may be taking us, and its mind-boggling implications—whether they occur in 2012 or some other time.

LIMITS TO CHANGE?

As explored in my 1992 book, *The White Hole in Time* (revised as *Waking Up in Time*), if the ever-accelerating pace of change continues, we are not going to be evolving for eons into the future. We could see the whole of our future evolution—as much development as we can conceive of, and more—compressed into a very short time. Within a few generations, perhaps within our own lifetimes, we could reach the end of our evolutionary journey.

It is often argued that this will never happen because there are limits to the rate of change. Any growth will eventually reach a plateau, resulting not in an ever-steeper curve, but in one that bends over into an S shape.

Population growth is a good example. For thousands of years, the human population has been growing and growing faster and faster. A thousand years ago, the world's population numbered around 310 million. This number had doubled by 1600. In 1800, it was approaching one billion, and the doubling time was down to 150 years. By 1960, it had reached four billion, with a doubling time of only 30 years. Since then, however, population growth has slowed; the curve has begun to bend over. If current trends continue, the human population will probably stabilize between 10 and 12 billion.

Similar S curves can be found in just about every area of development. For example, the production of steam locomotives

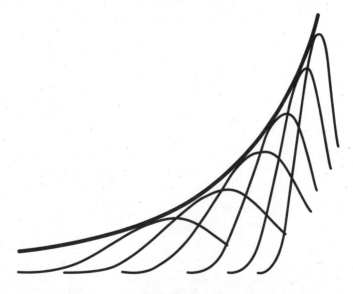

FIGURE 3: Successive S curves come faster, stacking up on each other. It is the overall appearance of the S curves that keeps increasing.

increased rapidly during the first century of the Industrial Revolution, then tapered off in the mid-twentieth century as diesel and electric power became more dominant. Or consider the growth of high-speed Internet connections in the United States. The rate of new connections grew rapidly in the first years of this century, and by 2005, over half of all homes had a high-speed connection. Now, as the saturation point approaches, the rate of growth of new connections has slowed.

However, when we talk about a speeding up of the overall rate of change, we are not talking of any particular S curve, but the rate at which successive S curves stack up. It took population growth thousands of years to reach its turning point. The industrial revolution took 200 years; high-speed Internet connections, less than a decade. So the question is not whether any particular growth keeps increasing forever, but whether there is a limit to the rate of ingression of novelty—whatever its medium at any particular time.

EVOLVING INTELLIGENCE

One recurring pattern that underlies evolution is an increasing complexity in the processing of information. DNA code is an information database, built up over eons. Sexual reproduction was an evolutionary breakthrough in information processing. So was the development of senses and, later, the central nervous system. The advent of human beings brought another major development in information processing—symbolic language—allowing us to share our thoughts and experiences with one another. Over the years, human breakthroughs in information technology—writing, printing, telephony, radio, television, computing, and the Internet—have consistently increased our ability to gather, process, organize, and utilize information.

Organization and utilization of information are the essence of intelligence. We usually think of intelligence primarily in human terms, and occasionally in other animals. But intelligence in its broadest sense has been evolving for billions of years. What is happening today with our own Information Revolution is but the latest phase of a process that has been going on since the birth of the universe.

So the question of whether there is a limit to the speed of evolution does not concern the limits of any particular phase of evolution; it is whether there is a limit to the rate of evolution of intelligence—whatever form it may take. As far as I can see, there is none.

BEYOND THE INFORMATION AGE

The growth of human information technologies is taking us rapidly toward a time when all human knowledge will be instantly available to anyone on the planet, in any medium. This will be a fully functional global brain in which the information technologies of television, telephone, and World Wide Web will be seamlessly integrated. The world's audio and video archives will be as easily accessible as text and images are today. Search engines will learn from their interactions with people, becoming increasingly sophisticated in their responses. We will be linked into an emerging global mind.

At this point, the growth rate of human knowledge will be reaching its own maximum. It, too, will begin to shift into an S curve. But knowledge is not the end point of the evolution of intelligence. Many have pointed to a hierarchy of data, information, knowledge, and wisdom. Information can be defined as the patterns extracted from raw data. Knowledge is the generalization of information to other situations. Wisdom determines how that knowledge is used. It involves discernment and evaluation: Is this decision for the better or worse? Will it help or hinder our future well-being?

At present, humanity has vast amounts of knowledge, but still very little wisdom. Without developing wisdom, it is most unlikely that we will avoid catastrophe. As the inventor-philosopher Buckminster Fuller repeatedly indicated, we are facing our final evolutionary exam. Is the human species fit to survive? Can we wake up sufficiently so that we can use our prodigious powers for the good of all and for that of many generations to come?

A HALF-AWAKE SPECIES

Symbolic language led to another significant step in human intelligence. We used language to communicate not only with each other, but also within our own minds—that is, verbal thinking. With this power, we could reflect upon our experiences and plan our futures. In addition, we could reflect upon the fact that we were aware. We became conscious of consciousness itself. We began to wake up to our own inner worlds.

At present, however, we are only half awake to who and what we really are. Becoming aware of our own selves brought with it a sense of an individual "I" observing the world and initiating our actions. But just what is this self? It seems so obvious that it is there, but, as many have discovered, it is hard to define or pin down.

When asked, Who are you? most of us will respond with the various things we identify with—our name, beliefs, occupation, education, roles, gender, social status, personality, interests. We derive a sense of identity from what we have or do in the world, our history, and our circumstances. But any such derived identity is conditional, and thus forever vulnerable. It is continually at the mercy of circumstances, and before long we find the need to defend or reassert our fragile sense of self. Our basic survival programming, designed to ensure our physical survival, is usurped by our psychological survival, leading to many unnecessary and often dysfunctional behaviors.

In addition, we are only half awake to our deeper needs and how to attain them. Most of us would like to avoid pain and suffering and find greater peace and happiness, but we believe that how we feel inside depends on external circumstances. This is true in some cases, for example, if we are suffering because we are cold or hungry. In the modern world, most of us can fulfill these demands very easily. The flick of a switch or a trip to the store usually suffices. But we apply the same thinking to everything else in life. We believe that if we could just get enough of the right things or experiences, we would finally be happy. This is the root of human greed, our love of money, our need to control events (and other people). It is the cause of much of our fear and anxiety; we worry whether events are going to be the way we think they should be if we are to be happy. This thinking is also at the heart of the many ways we mistreat, and often abuse, our planetary home.

The global crisis we are now facing is, at its root, a crisis of consciousness—a crisis born of the fact that we have prodigious technological powers but still remain half awake. We need to awaken to who we are and what we really want.

PROPHETS OF WISDOM

Throughout human history, there have been individuals who appear to have become fully awake. These are the enlightened ones—the mystics, seers, saints, rishis, roshis, and lamas who in one way or another have discovered for themselves the true nature of consciousness. Although their discoveries have been expressed in different ways, depending on the dominant worldview of their time, the essential message remains remarkably consistent. Aldous Huxley called this the "perennial philosophy"—the timeless wisdom that has been rediscovered again and again through the ages.

The enlightened ones have realized the illusory nature of the concept of a unique individual self. When we examine our experiences closely, delving deep into the nature of what we call "I," we find that there is nothing there—no *thing*, that is. This sense of "I-ness" that we all know so well, and that has been with us all our lives, is just our sense of being. It is awareness itself—so familiar, yet completely intangible. Thus, it cannot be "known" in the ordinary sense. Not realizing this, we seek to give our sense of self some form, some substance. We dress it up in various psychological clothes—all the things we think we are, or would like to think we are. This is the reverse of the emperor having no clothes. With true self-awareness, one discovers there are lots of clothes, but no emperor inside them.

Another consistent realization of the awakened ones is that the essential nature of mind, uncluttered by worry and chatter, is one of deep ease, joy, and love. Not recognizing this, most of us look to the world around us to provide us with peace and happiness. But despite all the messages from marketing and advertising industries, things and events do not bring happiness. On the contrary, our minds are so full of scheming, planning, and worrying whether or not we will get what we think will make us happy, we seldom experience the peace and ease that lie at our core.

When we awaken to our true nature, we are freed from a dependence on the external world both for our sense of self and our inner well-being. We become free to act with more intelligence and compassion, attending to the needs of the situation at hand rather than the needs of the ego. We can access the wisdom that lies deep within us all. This is the next step in evolution of intelligence: the transition from amassing knowledge to developing wisdom.

THE DAWNING OF A WISDOM AGE

Because each new phase of evolving intelligence takes place in a fraction of the time of the previous phase, we can expect the dawning of a Wisdom Age to take place in years rather than decades. It will be standing on the shoulders of the Information Age.

Never before have we been able to access so much spiritual wisdom. A century ago, the only spiritual tradition available to most people was the one that was indigenous to their own culture. Moreover, with rare exceptions, they did not have the benefit of learning from a truly enlightened being. Today we can access teachings from many different traditions and cultures, discover their common underlying truths, and translate that perennial philosophy into the language and terms of our own time. Something completely new is emerging: a single spiritual teaching that is a distillation of the world's wisdom traditions. This is coalescing and being disseminated globally through a variety of information technologies: books, tapes, Web pages, online forums, and Internet broadcasts.

At the same time, a growing number of people are becoming fully awake and proving themselves to be excellent teachers. Many are using the Internet to share their wisdom and help awaken others. Instruction in practices that facilitate awakening are already online and could become much more sophisticated. It may even turn out that *darshan*, the Indian word for a direct transfer of higher consciousness, can be transmitted via the net.

Awakening is often a sudden event. Once a person is ready—the necessary groundwork done, the circumstances propitious—the shift can happen more or less instantaneously.

It's possible that research into the neurological correlates of spiritual awakening will lead us to methods of promoting the process directly. There will likely be other unforeseen discoveries or developments that will help us free our minds. Whatever they may be, the more we learn how to facilitate a shift in consciousness, the faster it will happen.

As this becomes a mainstream phenomenon, humanity will relate to the world in wiser, more compassionate ways. Problems would still exist. Global warming would not suddenly cease; pollution would not evaporate; extinct species would not suddenly return. On the other hand, we might then have at our disposal new technologies that could help us solve the problems we have created. We can only guess at the ways in which this marriage of high technology and higher consciousness would play out. We have not been there before.

BEYOND WISDOM

Would this be the end point of our evolution? Or would there follow yet another turn of the spiral?

Many of the world's mystical traditions maintain that the liberation of the mind from its attachments is only the first step of inner awakening. More universal experiences of mind, and fundamentally different perspectives of reality, lie beyond.

Advanced adepts claim that the world of matter is not real and that space and time are not the ultimate reality. Interestingly, this view is in accord with modern physics' explorations into the nature of physical reality. Whenever we try to pin down the essence of matter, it eludes us. It seems nothing is there—that is, nothing of any material substance. Nor are space and time absolutes, as we once thought. They are part of a more fundamental reality, the space-time continuum.

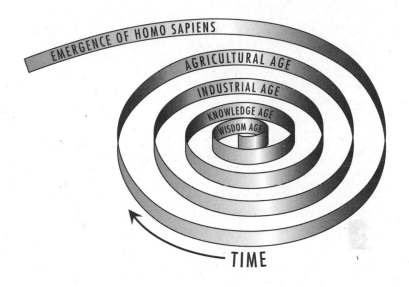

FIGURE 4: Successive turns of the spiral of evolving intelligence.

Perhaps those adepts have already discovered the ultimate nature of reality—not through digging deeper into its external forms, but through a penetrating exploration of inner space. If so, our collective destiny may be precisely this freedom from the illusion of materiality, from the illusion that we exist in space and time.

Let's not be too quick to rule out that possibility merely on the basis that it is so divorced from our current reality. If you had told Mozart that in the future humans would own tiny boxes the size of a coin, made from some strange material that was neither wood nor metal, with two strings coming out of the box that, when placed in their ears, would enable them to hear any of his compositions as clearly as if they were in a room with an orchestra, would he have believed you? On the contrary, he would probably have thought you mad.

THE OMEGA POINT

One person who believed our destiny was indeed a collective spiritual awakening was the French priest and paleontologist Pierre Teilhard de Chardin. Exploring the evolutionary trends toward greater complexity, connectivity, and consciousness, he argued that humanity was moving toward an "Omega Point"— the final end and goal of evolution.

He believed that the universe had been through several major stages of evolution, starting with what he called "cosmogenesis," the birth of the cosmosphere, the universe. Next was "geogenesis," the birth of the Earth (geosphere); following that, "biogenesis," the birth of life (the biosphere). With human beings, there came "noogenesis" and the "noosphere," the sphere of thought. He predicted that the final stage, the one that led to the Omega Point, would be "Christogenesis." This would be the birth of Christ consciousness, not in an individual but in the collective—the spiritual birth of humanity as a whole.

Teilhard de Chardin believed this Omega Point would happen thousands of years in the future. Like many others, he did not take into account the implications of ever-accelerating change. In his later years, he commented on the impact of radio and television in bringing humanity together. Technologies like these, he said, were bringing the Omega Point much closer. Just before he died, the first computers were being developed. Perceiving the potential of this new technology, he predicted that they, too, would bring the Omega Point even closer. If he had lived to see the emergence of the Internet, he would probably have realized that the Omega Point could come very soon indeed.

BREAKDOWN OR BREAKTHROUGH?

When we look at what is happening in the world today, it is understandable that we might scoff at the very idea of a collective spiritual breakthrough. The daily news may well lead us to believe that we are heading ever more rapidly toward breakdown rather than breakthrough.

That is, indeed, one likely possibility. I do not want to downplay the dire urgency of the world situation. If we don't make some radical changes, we are surely headed for disaster of one kind or another.

I also believe that positive change is possible. If we can develop the wisdom needed to navigate our way though these turbulent times safely, the potentials are staggering and unimaginable in scope. Let's put our hearts and minds to proving that we can pass Buckminster Fuller's final evolutionary exam and become a truly magnificent species. We are, after all, our only hope.

PART ONE

The Mayan Calendar and
Mayan Cosmovision

The Origins of the 2012 Revelation

JOHN MAJOR JENKINS

*The Mayan calendar is the source of the 2012 end-date, and
John Major Jenkins is one of its most well-known independent
researchers, having spent his entire career studying Mayan
cosmology and its meaning. Delving into the mythological
and spiritual connotations of the monuments at Izapa, an
ancient Mayan site, he reveals what the Maya foresaw
for the year 2012. In the following essay, John begins with
a "Mayan Calendar 101" primer. After addressing the basic
questions of Mayan philosophy, he guides the reader into a
deeper investigation of the spiritual teachings of 2012.*

When I think back to how I was introduced to 2012, I remember one warm summer afternoon when I was 12. I had a conversation with my childhood friend, Joe, that foreshadowed my lifework—my deep involvement as an adult with Mayan cosmology, eschatology, and end-date 2012.

Joe showed me a book his dad was reading about the Indians down in Mexico. They were astronomers and had a calendar that was going to end in the year 2011. "December 24, 2011," Joe said, and that date rang like a beckoning mystery in my mind. I chewed on it. *What could it mean? Would time just stop? Would the world blow up?* Curious, I looked at the book, a big hardcover with a dark blue dust jacket. I didn't fully register the author or the title, but 10 years later I had a deep feeling of recognition when I found Frank Waters's book *Mexico Mystique.* Yes, that was it. (Waters's book used an erroneous calculation and thus called the end-date December 24, 2011, but in truth, it is December 21, 2012.)

A chain of thoughts sprang up from that day with Joe, and something was triggered in my own cognitive development. The idea of "philosophy" was planted, the realm of ideas and complex thought, ancient cultures, history, and deep time. And yes, a specific seed thought was planted, that an Indian calendar from Mexico was going to end during my lifetime.

THE ANCIENT MAYAN STARGAZERS

To get to the root of the significance of 2012, it is important to understand the Maya and their calendar. The ancient Mayan day-keepers visited their mountain shrines and tracked the sacred sequence of days. But they were more than just counters. They were also stargazers, cosmologists, philosophers, and shamans. As shamans, they interacted with multiple dimensions to divine hidden secrets, access healing powers, and see beyond the veil of mundane appearances. Our appreciation of ancient Mayan genius must take into account this mystical domain of human endeavor, which today is often misunderstood and dismissed as primitive or superstitious. As with other New World societies, the Maya engaged in visionary shamanism as a precondition for the formulation of profound cosmologies and metaphysical teachings. Among the Quiché Maya, the visionary shamans were called the *nik wak'inel*—those who gaze into the cosmic center. According to them, the center of time and space would be revealed when their big calendar cycle came to an end—December 21, 2012.

How can we understand the deep and complicated legacy of the ancient Mayan stargazers? The approach I advocate for understanding 2012 and the Mayan calendar is self-evident but rarely practiced: let's base our understanding on the authentic Mayan documents that relate to the 2012 calendar and the World Age doctrine that is so intimately related to 2012. These documents are the Mayan Creation Myth (the *Popol Vuh*), the ball game mystery play that is central to the Creation Myth, and the carved monuments from the main ceremonial site of the early Mayan culture that invented the 2012 calendar—a site called Izapa.

By examining these core traditions, we can reconstruct and revive the original 2012 revelation. All religious traditions begin with a pure download, or revelation, from the transcendent source of all wisdom. Unfortunately, the original revelation always gets distorted and diluted during the ensuing centuries. Movements splinter, factions compete, corruption seeps in, and spiritual insights born of direct mystical experience get codified into religious dogmas designed to control access to spiritual truth. Worldly power plays co-opt and obscure the inner essence of the revelation. It would not be surprising if the original spiritual teachings of Izapa were likewise distorted as the Classic Period Mayan civilization grew to prominence. In fact, the increasing practice of warfare and human sacrifice at the end of the Classic Period (circa A.D. 900) is an indication of this decadent trend.

For this reason, it is extremely important that Izapa's carved monuments are preserved and are oriented to horizon astronomy in meaningful ways, so that we can decode what the creators of the 2012 calendar intended 2012 to mean. These monuments were built between 400 B.C. and A.D. 100. In studying them, instead of a quaint provincial belief system, we actually find a little doorway into the vast and profound perennial truths espoused by all of the great world religions at their deepest esoteric core. We find not only a profound astronomy-based cosmology, but also a prophecy and a spiritual teaching appropriate for all citizens of cycle endings. And the years leading up to 2012 signal a big cycle ending.

YOUR GUIDE

The keystone of the Mesoamerican calendar system is the 260-day tzolkin (pronounced zol-keen, derived from the Quiché Maya term *chol-qih*, or count of days). The tzolkin consists of 13 numbers and 20 day-signs. Each day-sign has an oracular meaning, with many layers of linguistic puns and metaphysical references that provide a rich database for Mayan calendar priests to weave their interpretations.

The tzolkin is a key that operates in many different domains of the universe. First and foremost, Mayan day-keepers (priests who count the sacred day-signs) offer a simple explanation for the tzolkin's magical properties. They say that the 260 days are based upon the nine-month period of human gestation. Mayan midwives, in fact, use the tzolkin cycle to estimate the birthdate of a child by adding 260 days to the day when a woman realizes she has missed her period. This reveals the tzolkin operating in the domain of human biology, human physical unfolding. It also suggests why significant life events often occur around one's tzolkin birthday: the primordial 260-day rhythm of embryogenesis continues as a resonant pattern throughout all of life's important junctures.

Second, the interval between the planting and harvesting of corn in highland Guatemala is 260 days. In the Mayan Creation

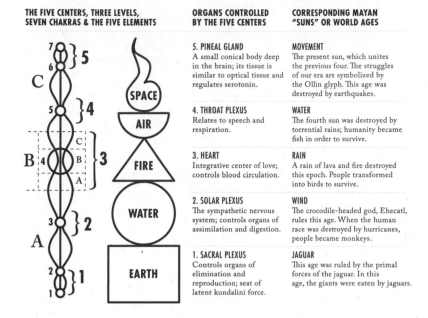

THE FIVE CENTERS, THREE LEVELS, SEVEN CHAKRAS & THE FIVE ELEMENTS	ORGANS CONTROLLED BY THE FIVE CENTERS	CORRESPONDING MAYAN "SUNS" OR WORLD AGES
	5. PINEAL GLAND A small conical body deep in the brain; its tissue is similar to optical tissue and regulates serotonin.	**MOVEMENT** The present sun, which unites the previous four. The struggles of our era are symbolized by the Ollin glyph. This age was destroyed by earthquakes.
	4. THROAT PLEXUS Relates to speech and respiration.	**WATER** The fourth sun was destroyed by torrential rains; humanity became fish in order to survive.
	3. HEART Integrative center of love; controls blood circulation.	**RAIN** A rain of lava and fire destroyed this epoch. People transformed into birds to survive.
	2. SOLAR PLEXUS The sympathetic nervous system; controls organs of assimilation and digestion.	**WIND** The crocodile-headed god, Ehecatl, rules this age. When the human race was destroyed by hurricanes, people became monkeys.
	1. SACRAL PLEXUS Controls organs of elimination and reproduction; seat of latent kundalini force.	**JAGUAR** This age was ruled by the primal forces of the jaguar. In this age, the giants were eaten by jaguars.

FIGURE 5: Time, space, spirit, and matter united in Mayan cosmology
(from *Journey to the Mayan Underworld*, John Major Jenkins).

Myth, human beings were made from corn dough, so this is a reflection of human gestation on an agricultural level. Third, the tzolkin is the critical key to the Mayan almanacs that predict eclipses and the movements of the sun, moon, and planets. The day-signs and numbers of the tzolkin provide a calendrical framework with which to track celestial movements. For example, the Sacred Day of Venus, which inaugurates the 104-year Venus Round, always occurs on the tzolkin day 1 Ahau.

With these three examples, it is easy to see how the tzolkin is a key to human cycles, earth cycles, and celestial cycles. The tzolkin unites the processes of heaven and earth and is the key to a mystical philosophy of "as above, so below" and "the microcosm reflects the macrocosm." The Maya formulated a grand comprehensive vision of universal processes united by a common numerical thread woven throughout the tapestry of multiple dimensions.

The Maya also developed a timekeeping system called the "Long Count." This is the calendar that targets 2012 as the end of a vast cycle of time, a key concept in the Mayan doctrine of world ages. The Long Count utilizes five place values:

1 day = 1 kin (day)
20 days = 1 uinal (vague month)
360 days = 1 tun (vague year)
7,200 days = 1 katun (19.7 years)
144,000 days = 1 baktun (394.26 years)

Thirteen baktuns equal 1,872,000 days, which is one World Age cycle (5,125.36 years). Notice that to reach the World Age cycle, the baktun period is multiplied by 13 rather than 20. We know that the 13-baktun period ends on December 21, 2012, because scholars have deciphered how the Mayan calendar correlates with our Gregorian calendar.

Modern scholars use a simple Long Count notation. For example, 9.16.4.4.1 means that 9 baktuns, 16 katuns, 4 tuns, 4 uinals, and 1 day have elapsed since the zero day of the Long Count, which the correct correlation fixes at August 11, 3114 B.C. The "zero" date is written 0.0.0.0.0, and the "end-date" of the 13-baktun cycle is thus written 13.0.0.0.0. The use of the term "end-date" gives rise to the mistaken notion that the Mayan calendar ends in 2012. Mayan time is cyclic, and it should go without saying that time continues into the next cycle. We use similar conventions in our language when we speak of "the end of the day," but we don't expect time to stop at midnight.

How do we know December 21, 2012, is the correct end-date? Why not A.D. 2230 or A.D. 1740? This issue involves the correlation of the Mayan calendar with our own, an important question that was of great concern to early Mayan scholars. After decades of research, the correlation question was finally settled in 1950. An unexpected result of settling the correlation was that the thirteenth baktun would end on December 21, 2012, on the tzolkin day 4 Ahau. This date in the tzolkin confirmed the surviving tzolkin day-count in highland Guatemala and validates the carvings called "creation monuments," which always correlate 13.0.0.0.0 with 4 Ahau. A fundamental fact that needs to be appreciated, so that we may all act in unity, is what I call "the equation of Mayan time":

$$13.0.0.0.0 = \text{December 21, 2012} = 4 \text{ Ahau}$$

This "equation" shows how the tzolkin, the Long Count, and the Gregorian calendar are connected.

Although Mayan time-philosophy is a deep study that can get complex, for the purpose of understanding 2012 we only need to know that December 21, 2012, is the end of the 13-baktun cycle in the Long Count calendar, which in Mayan philosophy equals one World Age. Despite misunderstandings vigorously repeated by critics, the 2012 end-date is firmly established and is a true and accurate artifact of the Mayan philosophy of time.

MAYAN METAPHYSICS AND SPIRITUAL COSMOLOGY

Having completed the "Mayan Calendar 101" primer, we can now explore more interesting territory. This is where we open our minds to the profound metaphysical truths that hide within the Mayan calendar. Philosophy, spirituality, metaphysics, cosmology, religious rituals, perennial wisdom, and time-tested teachings about transformation all lie within the domain of the Mayan calendar. What emerges from studying Mayan time-philosophy is nothing less than a grand galactic cosmovision unparalleled in world traditions. Yet the Mayan tradition, at its core, advocates and elucidates the universal and perennial wisdom teachings at the heart of all the world's great spiritual traditions.

My own journey with the Mayan calendar has been like a deepening initiatory path. When my journey began, I had a dream that I was an old man speaking to youngsters. When they asked who I was, I chuckled and replied, "I am, and will always be, a student of Mayan time." This is an ever-deepening path. The breakthrough occurs when we see that Mayan spiritual teachings are not the arbitrary beliefs of an obscure people, but teachings that tap into the great perennial truths that all spiritual traditions share. But there's something more about the Mayan tradition—it exemplifies a great scientific and spiritual achievement in the Americas.

As we explore Mayan time-philosophy, our guiding light will be the expectation that profound perennial wisdom is waiting to be recovered. The Mayan calendar embodies a comprehensive cosmovision, a grand unified theory of time, space, spirit, matter, and consciousness. The Maya were much closer to the perennial metaphysics known to students of esoteric teachings and oriental mysticism, a kind of high-minded New World shamanism firmly rooted in good math and astronomy. Calendar, myth, astronomy, and spiritual awakening: all these themes were woven together in the Mayan calendar.

COSMIC CENTERS AND GALACTIC ALIGNMENTS

The Mesoamericans were very interested in orienting themselves correctly to the larger universe, and this was expressed in their architecture and city building plans. It was also the centerpiece of their cosmology and religion. They were, after all, the *nik wak'inel*—those who gaze into the center. But where is the center?

The ancient Olmec civilization that preceded the Maya believed the center was the polestar, around which all the other stars appear to rotate. Other traditions, such as the New Fire ceremony, utilized the zenith (the exact center of the sky overhead). Along with the polar center and the zenith center, the Maya acknowledged a third cosmic center hidden within the 2012 end-date and the World Age doctrine preserved in the Mayan Creation Myth.

The Milky Way crosses over the ecliptic (the zodiacal path followed by the sun, moon, and planets) in two places—one in Sagittarius and one in Gemini (see Figure 6). In Mayan symbolism, crosses denote the idea of "cosmic center." What is interesting about these established Mayan concepts is that the cross in Sagittarius targets the center of our Milky Way galaxy (the galactic center).

Identifying the center of the world was clearly an essential concern for the Maya. But there were unavoidably three answers, or levels, to this question. No Mayan scholars and few others have been willing to accept this slight complication, but the payoff is that we gain an accurate understanding of Mayan spiritual philosophy. Since the early 1990s, I've explored the various traditions around the three cosmic centers. The astronomy aside, the Maya also saw them as coequal principles associated with three deities: Seven Macaw (Big Dipper, the polar center); Quetzalcoatl (the Pleiades, the zenith center); and One Hunahpu (December solstice sun, the galactic center).

But how is the December solstice sun associated with the galactic center, not to mention 2012? On this third level, the galactic center emerges as the senior cosmic center, the one that embraces the highest possible considerations and the most comprehensive perspective. In the same way that the heliocentric model introduced by Copernicus in the sixteenth century was a superior way of modeling the cosmos, the Maya achieved a cosmological understanding that superceded earlier and less true perspectives. So, how is the December solstice sun deity (One Hunahpu) associated with the galactic center? Here is the key to 2012: in the years around 2012, the December solstice sun will align with the center of the cross that targets the galactic center.

This era-2012 alignment is caused by the phenomenon known as the precession of the equinoxes. The Earth wobbles very slowly on its axis and changes our orientation to the larger field of stars, including the Milky Way. For the ancient sky watchers, the most noticeable effect of this phenomenon was that the position of the sun on the equinox shifted in relation to background stars and constellations (as well as the Milky Way). The phenomenon also equally affects the solstices, so the position of the December solstice sun has been shifting slowly, appearing to converge with the Milky Way's center over many thousands of years. In fact, the December solstice sun aligns with the Mayan cross and the galactic center only once every 26,000 years—the full cycle of precession. This is an apparent alignment that was meaningful to the ancient stargazers.

My research into the empirical basis of this event reveals that an era of alignment must be considered—an alignment zone for era-2012 that stretches from 1980 to 2016. When they formulated the Long Count calendar, the ancient Mayan astronomers calculated forward to the alignment and fixed the end of the 13-baktun cycle to it.

This is the hidden astronomical basis of the 2012 end-date. Other Mayan traditions utilize the alignment image as

FIGURE 6: The Mayan sacred tree and the galactic alignment in era-2012 as viewed from the Earth: the December solstice sun converges with the galactic center over thousands of years.

a primary motif. For example, I discovered that the sacred ball game and the Mayan Creation Myth encode the 2012 alignment, the "galactic alignment." Similarly, I discovered that Izapa was where the 2012 calendar originated and is a rich source of carvings, prophecy, and spiritual teachings that relate directly to 2012. No one else had taken this approach, and my findings promised to turbocharge Mayan studies.

My investigations and discoveries took place at a rare juncture in my life. By 1994, I'd lived and worked in Mayan villages in the highlands of Guatemala, had visited many of the temple sites throughout Mesoamerica, and was engaged in an ongoing effort to push back the fringes of what was known about ancient Mayan cosmology. With the galactic alignment, I felt I had found the key to understanding 2012. I made a conscious decision to dedicate myself to studying and analyzing the implications of what I was uncovering. Living in a renovated garage outside of Boulder, Colorado, I worked part-time to pay my bills and wrote dozens of essays, each one blazing a new trail in explaining how the galactic alignment was encoded into Mayan traditions. The nearby university library was my research archive.

In 1998, when I completed my book *Maya Cosmogenesis 2012*, I was confident that my findings were supported by the evidence—that I had put a key concept on the table that made sense of so many disconnected threads in how official scholars interpreted Mayan traditions. The very fact that the 13-baktun end-date fell on a solstice date highlighted some kind of intention behind it, but scholars dismissed this as coincidence. I was bemused and baffled at how any progress could take place in academic circles when that kind of attitude runs the show. Yet, I was following the trails laid out by the scholars themselves. Michael Coe, for example, had said that "the priority of Izapa in the very important adoption of the Long Count is quite clear cut."[1]

The Izapan monuments confirmed that I was on the right track and was asking the right questions. The three main monument groups at Izapa refer to the three cosmic centers and their associated avatars. A tripartite sacred science was pioneered here. The Izapan ball court is the monument group that refers to the galactic alignment and the galactic center—One Hunahpu's resurrection. The ball court also contains the carved monuments that explicate the 2012 prophecy and the spiritual teachings that apply to cycle endings.

The end-date alignment can be thought of as an eclipse, and it shares with eclipses the basic alchemical meaning of "the transcendence of opposites." In Mayan metaphysics, this union has a more profound meaning that goes beyond the union of male and female and other pairs of opposites. Instead, it involves the nondual relationship between infinity and finitude, eternity and time—a union of lower and higher. Our higher and lower natures are reunited in eclipses, in the Quetzalcoatl myth of the sun joined with Venus, and in the 2012 eclipse of the galactic center by the solstice sun. Our higher nature does not destroy our lower nature, but embraces it. Likewise, time is restored to its relationship with timelessness when human consciousness reclaims the timeless, eternal perspective. The manifest world of appearances is restored to limitless possibilities when human consciousness reestablishes its connection to infinity.

This is the heart of the promise of world renewal in 2012, and it can only take place within the heart of humanity. The indispensable secret ingredient in making these possibilities manifest is sacrifice, or surrender. We must surrender our ego-driven fixation to states of limitation, surrender the illusory ties that occlude our direct perception of eternity and infinity. We do not evolve toward these states because they reside at the root of our beings; instead, we unveil them by letting go of the limitations that obscure the reality of their immanent presence.

These principles of Mayan sacred science are really perennial truths. They are found again and again on the carved monuments of Izapa. It is time we take a closer look at this amazing and underappreciated site, a place I characterize as the New World Eleusis, which harbors unrecognized Rosetta stones and forgotten initiatory visions.

Although it is not possible to do Izapa justice in a brief article, a general survey can at least provide the framework for
understanding the revolutionary cosmovision pioneered at this
humble, profound fountainhead of Mayan wisdom. If understanding 2012 from the vantage point of the creators of the
2012 calendar has any meaning at all, I can't imagine discussing
2012 without reference to Izapa. The important points are:

- The three cosmic centers and their avatars.
- The ball court monument group encodes the galactic
  center alignment and the 2012 prophecy—it is "ground
  zero" of the 2012 revelation.
- The ball game and the Creation Myth encode the
  astronomy of the galactic alignment in era-2012.
- Izapa is an initiation center where practical methods can
  be found for spiritual seekers to directly understand what
  2012 means.
- Izapa's 2012 revelation is an expression of universal
  perennial wisdom.

The "Izapan civilization" thrived between the Olmec and the
Maya, 450 B.C. to A.D. 50. Today, a dozen or so archaeological sites preserve monuments that are stylistically connected to
Izapa, which was clearly the most important ceremonial site in
the region. The sun passes directly over Izapa, located between
the tropics at 15 degrees latitude, on May 1 and August 11. This
fortunate circumstance divides the year into 260-day and 105-
day sections. For this reason, some scholars believe that Izapa is
where the 260-day tzolkin calendar, as well as the Long Count,
originated. Izapa's geodetic position is uniquely appropriate,
because at other latitudes the division intervals are different.

The Long Count calendar first appears in archaeological
records in the first century B.C., precisely during Izapa's heyday.

The earliest Long Count monuments are distributed in a wide arc around Izapa, and a most compelling one is located in Izapa's sister city, Takalik Abaj, dated to A.D. 37.

Izapa was excavated in the 1960s by scholars from Brigham Young University. They identified over 60 carved monuments, many of them found exactly as they were left when the site was abandoned some 1,800 years ago. For this reason, the horizon orientation of the monuments and the main groups can be charted. My book *Maya Cosmogenesis 2012* broke the case on Izapa and showed how the deities and mythological episodes depicted on the carvings encode and reflect the astronomical movements over Izapa. This is not at all surprising, since we now know that Mayan mythology and astronomy are intimately related. The mythological scenes are recognizable—they involve the Hero Twins, the vain ruler Seven Macaw, and the solar father of the twins, One Hunahpu. It thus appears that the Mayan Creation Myth is a central theme of the Izapan monuments. In fact, some of the earliest Creation Myth themes known in the archaeological record are found at Izapa, suggesting that along with the Long Count, the Creation Myth also originated here.

The three main monument groups at Izapa—Groups A, B, and F—represent the three cosmic centers identified in Mayan astronomy: polar, zenith, and galactic. This is clear from the groups' orientations as well as the carvings' symbolic content. More striking are the portrayals of the three well-known deities associated with the three cosmic centers, engaged in various triumphs and exploits: Seven Macaw, Quetzalcoatl, and One Hunahpu. Seven Macaw is a bird deity in Group A who represents the Big Dipper constellation that circles the polestar in the northern sky. Another bird deity at Izapa is probably an early version of Quetzalcoatl, who is shown ascending into the zenith on one of the monuments in Group B. We glimpse here a complex mystery play of three cosmic forces that illustrates spiritual lessons for humanity, especially in regard to the dynamics that unfold during cycle endings. The entire Creation Myth is a World Age doctrine concerned with vast eras and their transition points, and it thus speaks directly to our 2012 questions.

In order to understand the prophecy and the teachings for 2012 preserved in the Group F ball court, we can focus on the action between Seven Macaw and One Hunahpu. These stories are not quaint fairy tales. They encode profound perennial wisdom, spiritual teachings for citizens of cycle endings. Figure 7 illustrates two carvings showing Seven Macaw ascending and defeated.

These carvings are oriented toward the north, toward Tacaná Volcano, over which the Big Dipper rises and falls. The carvings of Seven Macaw illustrate a very human lesson of ambition, hubris, and demise. He is the vain and false ruler of the previous World Age, hellbent on deceiving humanity. His earthly power comes from a total commitment to self-serving egoism. But this power comes at a price, for his devotion to such a one-sided worldview cuts him off from knowing his true creator, his infinite and eternal source. He suffers in his alienation and projects his pain onto the world.

The Hero Twins enter the scene after their father, One Hunahpu, gets his head cut off by the Lords of the Underworld. Their destiny is to avenge their father's death, facilitate his rebirth, and reinstate him as the new World Age ruler. But they must first do away with Seven Macaw. The illustration on the left in Figure 7 depicts the scene of a Hero Twin shooting at Seven Macaw, causing him to fall from his perch, his throne in the polar center of the northern sky.

POLESTAR

FIGURE 7: Izapan monuments showing the Big Dipper rising and falling over Tacaná Volcano.

The incredible carving illustrated by Figure 8—a cosmological New World Rosetta stone—resides in a back room of the museum in Tapachula, a town about 10 miles from Izapa. Notice the bird in the tree—Seven Macaw. Hunahpu, the son of One Hunahpu, wants to do away with Seven Macaw. The alligator on the left represents the Milky Way, and its head at the bottom is the nuclear bulge of the galactic center. Notice the nice up-down polarity that connects the mouth of the alligator to the polar bird deity above. An amazing knowledge is encoded in this image, an incredible awareness of the north-south dialectic between the polar center and the galactic center. The astronomical players define the mythological action, which unfolds as a mystery play intended to teach a spiritual lesson about cycle endings—about 2012.

The alligator's mouth is an important feature in Mayan mythology. Corresponding to a "dark rift" caused by interstellar

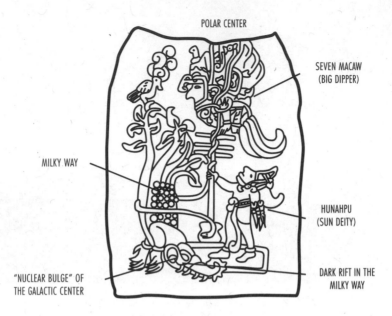

POLAR CENTER

SEVEN MACAW
(BIG DIPPER)

MILKY WAY

HUNAHPU
(SUN DEITY)

"NUCLEAR BULGE" OF
THE GALACTIC CENTER

DARK RIFT IN THE
MILKY WAY

FIGURE 8: Stela 25: Izapa's cosmological Rosetta stone shows the Milky Way, galactic center,
Big Dipper, and sun deity arranged according to the polar and galactic opposition.

dust that runs north along the Milky Way from the nuclear
bulge of the galactic center, the Maya called it the *Xibalba be*
(the road to the otherworld). The alligator's mouth is also portrayed as a frog's mouth on several monuments, including the
amazing Stela 11 (see Figure 9), which contains unambiguous
motifs of a galactic mythology.

In Mayan hieroglyphic writing, the upturned frog-mouth
glyph means "to be born." The deity emerging from the frog's
mouth is One Hunahpu, the primal First Solar Lord who measures the cosmos with his outstretched arms, at the dawn of the
new age. He has calendrical, directional, and seasonal associations that identify him with the December solstice sun. Thus,
Stela 11 is an image of the December solstice sun being reborn
from the dark rift. This is, in fact, a picture of the rare galactic

FIGURE 9: The author with Stela 11, an iconic statement that says, "The December solstice sun is reborn from the dark rift in the Milky Way." This is the rare galactic alignment of era-2012.

alignment that culminates in era-2012. Let me emphasize this profound discovery: Stela 11, though carved over 2,000 years ago, depicts the rare astronomical alignment that occurs once every 26,000 years and that culminates in era-2012. Because the road to the otherworld is associated with ceremonial caves, temple doorways, and the human birth canal, we can understand how the Maya thought of the 2012 alignment as a sacred rebirth. Izapa's 2,000-year-old iconographic code makes a striking statement: the world is reborn when the December solstice sun aligns with the dark rift in the galactic center.

As if this weren't enough to place Izapa on the short list of World Heritage sites, in Izapa's monument site known as the ball court, the galactic alignment of era-2012 is integrated with a mystery play that conveys a spiritual teaching. Here the astronomy, the prophecy, and the spiritual teaching for 2012 are one. This coherent integration underscores the absolutely incredible and indispensable role that Izapa holds for understanding 2012.

Let's tune into the ball court (see Figure 10 on page 59), ground zero of the 2012 prophecy. First, it is important to understand how the ball game symbolizes the galactic alignment. Drawing from accepted symbolic connections between astronomy and the ball game's iconography, we can say a few things with certainty. The ball court represents the Milky Way. The game ball represents One Hunahpu's severed head—that is, the December solstice sun. The goal ring represents the dark rift in the Milky Way.

The ball players act out the ball game/mystery play recorded in the Popol Vuh Creation Myth, wherein the Hero Twins battle the Dark Lords of the Underworld in the cosmic ball field of action, sacrifice, justice, and resurrection. We can see that the ball game itself is the centerpiece of the Creation Myth's encryption of the galactic alignment of era-2012. When the ball goes into the goal ring, the solstice sun goes into the dark rift, and the cycle of time is over. At that moment, the old World Age of delusion and deception draws to a close, and a consciousness renewed and elevated by rebirth in a higher light takes center stage.

The monuments in the Izapan ball court preserve a coherent message about this mystery play that came to the Izapan sky watchers as a revelation from a higher source. The 2012 message needs to be understood in the context of the ball court's orientation to the horizon of the December solstice sunrise, the role of the throne-sitting shaman, the true meaning of sacrifice, the energy dynamics of cycle endings (in era-2012), and the relationship between ego and the divine self. Let's take on this last point first.

The Mayan Creation Myth of Seven Macaw and One Hunahpu parallels a perennial teaching that was best summarized by Aldous Huxley in his introduction to *The Song of God: Bhagavad-Gita*. Distilling the universal wisdom found at the core of all of the world's great religions, Huxley described four principles. First, human beings contain two selves—the limited ego-self, which serves a purpose but can give rise to megalomania, and the divine self, which is the eternal and infinite ground of all manifestation. The world, including its objects, elements, stars, planets, and various phenomena (including living beings), all spring from this unconditioned divine ground. Second, overidentification with ego gives rise to a forgetting of the divine self. Third, the unlimited divine self cannot be conceived or understood by the rational intellect, but it can be directly experienced. This experience comprises the mystical initiation spoken of in the esoteric teachings of all religions. Fourth, the purpose of human existence is to reunite the ego with the divine self and live life in full consciousness of the eternal and infinite ground of all manifestation.[2]

The key here, for our times, is the initiatory glimpse that is possible and that can give meaning and purpose to life in the modern world. How is this done? The monumental message of Izapa points the way with three methods: tantra, shamanic healing with sacred plants, and breathing meditation. The key to the efficacy of all three is sacrifice (or surrender). And the key to sacrifice is humility and compassion.

Figure 10 on page 59 represents the ball court at Izapa and its various monuments. It is oriented toward the southeast horizon where, in 2012, the December solstice sun will rise in union with the dark rift and the nuclear bulge of the galactic center. Some 2,100 years ago, the Izapan astronomers would have viewed the dark rift rising above this horizon in the predawn skies of a December solstice. Although a separation of some 30 degrees was evident back then, precession has caused the sun and the galaxy to converge and appear now—from the Earth's vantage point—to align. The ancient cosmologists observed and calibrated this convergence and calculated its occurrence in 2012, anchoring their calendar and end-date cosmology to it. (There are reasons for believing that this alignment bodes a true energy dynamic in the cycles of time, as I explored in my 2002 book, *Galactic Alignment*.)

On the far side of the ball court, a monument depicts Hunahpu standing over a fallen Seven Macaw. Viewing this monument with an awareness of the future alignment that will occur behind it, the Creation Myth story was reiterated for the ancient initiate: Seven Macaw (the ego) must fall before One Hunahpu (the divine self) can be reborn. The old polar god dies, and the new galactic god is born. A solar-galaxy model supercedes a geo-polar one. Ego must be humbled before it can be restored to right relationship to the divine self, allowing the eternal wisdom to shine through.

This combination of astronomy, mythology, and spiritual teaching encodes what I identify as the 2012 prophecy. The scenario applies to the transition point at the end of the age. What did the Maya prophesy? Seven Macaw, self-serving egoism, will be ruling and ruining the world, tricking people into forgetting their eternal natures and identifying with their most limited selves. That seems the truest prophecy I've ever heard for 2012. The second part of the prophecy, also nicely portrayed in the Creation Myth, is that ego *can* humble itself, surrender its illusory belief that it is the true center, and reconnect with the eternal divine self. In this way, One Hunahpu, the soul reborn in right relationship with the higher source, can shine through. Ego grows transparent and becomes illuminated by the divine mind that was hidden

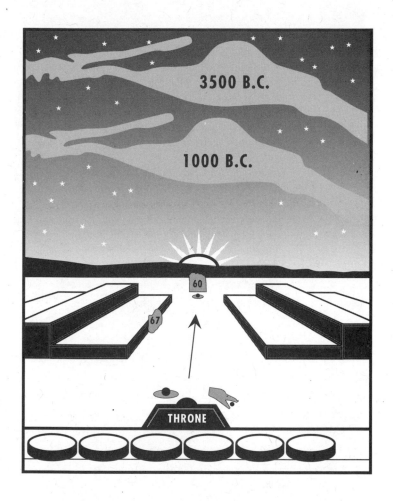

FIGURE 10: Izapa's ball court monuments encode the galactic alignment of era-2012, as well as serving as a prophecy and a spiritual teaching.

underneath. The key to this prophecy is free will, for what we actually experience depends on our choices—whether we cling in fear to familiar illusion or open up in love and trust to the great mystery that seeks to transform and elevate us.

FIGURE 11: Stela 67: The December solstice sun deity, reborn from the Underworld
at the end of the age, in the middle of the Milky Way canoe.

In the center of the north wall of the ball court we find Stela
67, showing One Hunahpu's victorious rebirth in the center of
a canoe that symbolizes the Milky Way. As in Stela 11, he is
shown in the galactic alignment position, in the "middle" of
the Milky Way. A viewer of this monument could look above
and beyond it to the north, to Tacaná Volcano, where Seven
Macaw rises and falls, a reminder of the vicissitudes and ultimate folly of egoism.

On the near end of the ball court, there are six council seats,
or observation thrones (see Figure 10). A delegation of teachers
or shaman sky watchers sat on these, observing the mystery play
unfold on the court, as well as the skies over the far horizon.
They could look over the ball court's throne, on which the principal initiate sat facing the December solstice horizon. Let's take
a look at this throne and the nearby monuments (see Figure 12).

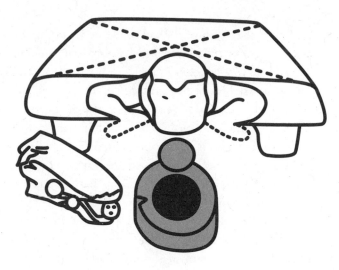

FIGURE 12: Izapa's 2012 prophecy: Three iconographic statements encode the galactic alignment and offer a message of transformation, rebirth, and free will.

These three sculptural statements—the throne, the game ball, and the serpent's mouth—encode a three-part teaching about 2012, and all of them symbolize the alignment of the solstice sun with the dark rift. First, Mayan thrones represent the center of the cosmos. The throne shows a solar godhead emerging from between stylized legs. Clearly, sun and birth canal are portrayed, which reflect the future alignment to occur over the far horizon, the December solstice horizon. This is an image of rebirth. Second, the serpent head on the lower left (which originally held a little solar head in its mouth) is another way of depicting the sun in the dark rift. But this has the connotation of death, of devouring.

So far, we have a death and rebirth, a polarity that is often used to depict the December solstice, which marks the end of one year and the beginning of another. A new birth must be preceded by the death of the old. But the devouring image also means transformation. That which dies doesn't end, but is transformed into that which is reborn. Likewise, ego isn't to be annihilated, but transformed by contact with the renewing light of the divine self. This is the true meaning of sacrifice and transcendence, for transcendence does not annihilate that which it transcends, but includes it in a more comprehensive vision.

The stone game ball and ring in the lower part of the figure represent the ball game and, as we have already seen, the galactic alignment of the solstice sun and the dark rift. The meaning here involves the game, the play of consciousness, the dance, the activity of humans moving toward rebirth, and conscious participation in the process. This relates to a requirement of positive transformation, that of a *willing* surrender of the illusions that bind ego to states of limitation. Seven Macaw refused to surrender his attachment to ego, and so falls in the myth. One Hunahpu willingly went to Xibalba and lost his head, but gained it back, and so much more, when he was reborn.

The three methods of surrender and reconnection to the divine self were clearly employed in Izapa's esoteric religion. The first is a form of tantric teaching, evident in the conceptualization of the alignment as a union of cosmic father and cosmic mother. These partner principles represent two cosmic domains united: the solar and the galactic—father sun and mother galaxy. The sexual polarity of the symbolism suggests that some form of tantric ideology was known at Izapa.

The second method is shamanic use of sacred plants, an extreme technique of soul retrieval, healing, and temporary circumvention of the control systems of ego, the goal being transcendence of limitation and restoration of the whole self. This method for traveling to higher dimensions has been employed

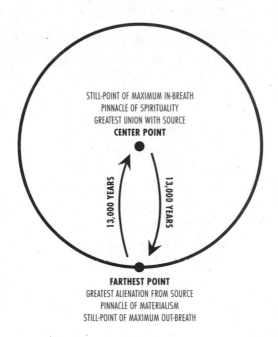

STILL-POINT OF MAXIMUM IN-BREATH
PINNACLE OF SPIRITUALITY
GREATEST UNION WITH SOURCE
CENTER POINT

13,000 YEARS

13,000 YEARS

FARTHEST POINT
GREATEST ALIENATION FROM SOURCE
PINNACLE OF MATERIALISM
STILL-POINT OF MAXIMUM OUT-BREATH

FIGURE 13: 2012 is the center of Mayan time. The still-point of the out-breath is analogous to the era-2012 alignment and can provide a glimpse of the eternal transcendent wisdom.

by shamans for millennia. Ritual mushroom stones found in the region of Izapa suggest that psilocybin mushrooms were used shamanically to produce altered states and visions.

The third method is suggested by the ball court throne and is the most accessible to modern seekers wishing to experience for themselves "what 2012 means." Many Mayan paintings and carvings depict shamans or kings sitting on thrones in the classic full lotus posture. This posture is associated with deep absorption into a mystical experience of timeless unity, a glimpse of eternity. The ancient Mayan initiates must have employed some form of meditation practice, much like their counterparts in Asia. Essential to this practice is breathing practice, or *pranayama*.

Here, I will take a small leap and propose that Mayan mystics practiced a yoga and breathing meditation, which in Asia is called Vipassana Meditation. This practice focuses on the breath cycle. In this meditation practice the still-point of maximum out-breath is a doorway beyond the vicissitudes of time's endless cycling, and there the seeker can experience the still quietude of eternity. Vipassana is perfectly suited to be the preeminent spiritual teaching for those who wish to experience what 2012 means, because the breath cycle maps onto the peaks and troughs of natural cycles, including the great 26,000-year cycle of precession that is coming to completion with the galactic alignment in era-2012. This is how it works:

How do we know that the circle of precession ends in 2012? A cycle is a circle, so how can we identify its beginning and end points? The breath cycle provides a clue. Cycles in nature follow movements of increase and decrease. The day cycle has two phases—increasing light and increasing darkness. The moon waxes and wanes, because it appears to move closer to the sun and then farther away from the sun. The period of "moon dark" is the new moon period, when the sun and moon are conjunct. An eclipse is a precise conjunction of sun and moon, and in alchemy, eclipses symbolize the union of opposites, the reunion of the moon with its light source.

Likewise, precession can be divided into eras, or World Ages, using the 12 signs of the zodiac. But what really defines the energetic divisions of the precessional Great Year? The key to the precessional cycle's beginning and end points is the galactic alignment. It makes perfect sense to think of the December solstice sun as the hour hand of the precessional clock, and when it comes around to touch the galactic center, galactic midnight will occur. The end of the cycle, galactic midnight, happens when the solstice anchor of the seasonal shifting strikes the galactic root of time and space. Time, in this conception, moves in and out of union with the source. End-date 2012 is not the end of circular time, but the center of time's breathing in and out.

Human civilizations experience this in-out movement over thousands of years, like the breathing of history. From the Golden Age of union with source to the Iron Age of abject materialism and alienation from source, we move from the noon of light to the midnight of darkness. Midnight is the still-point between death and rebirth, and the alignment point corresponds to the maximum out-breath. Mapped onto the Earth's breathing cycle, which we could think of as the 26,000-year precessional cycle, the still-point is the galactic alignment of era-2012.

In Vipassana Meditation, one enters into this still-point and opens into a timeless space. The center and source of time can open for us in 2012. The key is found in our own breathing, our inner spiritual link to our divine selves that must be reopened. The throne at Izapa invites us to sit in meditation and enter the still-point alignment of the male and female currents, ascending the ladder of vision to the spiritual plateau where the vicissitudes of time are transcended. In that place, all of the issues and conflicts that seem so intractable when we are stuck in lower planes of awareness are revealed for what they are: illusions. Illusions that gave power to the limited ego to the extent that we believed in them.

And here's the nub of the sacrifice doctrine: our goal is not the sacrifice of literal hearts, bodies, or even our egos, but the sacrifice of our illusions. In that surrender, we lose nothing real but gain everything worth having: our true selves. From this perspective, the "prophecy" depends on how we respond to the moment, the eternal now. As 2012ologist Geoff Stray put the essential question, "2012, is it catastrophe or ecstasy?" It seems to me that apocalyptic catastrophes will be true to the extent that we cling to our illusions. The divine flame of ecstasy will flare in direct proportion to the illusions tossed into it. Feed the ecstatic flame of the divine mind, for all true knowledge comes from an ecstatic connection with the transcendent.

BACK TO THE CENTER

Izapa, the origin of the 2012 calendar, is the New World Eleusis, a pioneering initiatory center into galactic wisdom. If we contemplate the carvings and the orientations of the monument groups and understand the symbolic language, a profound and unrecognized paradigm is revealed, and the package is complete. Not only do we find the hard science of a galactic cosmology anchored to a rare astronomical alignment in era-2012, but we also find a creation myth, a prophecy, spiritual teachings, and techniques with which citizens of the cycle ending can experience for themselves what 2012 means.

A high-order revelation occurred at the dawn of Mayan civilization. As with all revelations, it was the discovery of undying perennial wisdom—the same wisdom that is found at the root of all of the world's great spiritual traditions. The pure and undiluted revelations that spring forth at the dawn of a civilization or a religion have a way of being polluted as time passes, getting obscured as religion becomes dogmatic and hierarchical. We are blessed that the carved monuments of Izapa were found undisturbed, silent sentinels to a grand cosmic religion that saw the galactic center as the great transformer and the galactic alignment of era-2012 as a great opportunity for spiritual seekers to reconnect with the source of perennial wisdom. But we shouldn't wait until 2012, for we are in the alignment zone now (1980–2016). The time, as always, is now.

The Mayan Factor
PATH BEYOND TECHNOLOGY

JOSÉ ARGÜELLES, PH.D.

One of the popular leaders in the field of modern Mayan calendar predictions, José Argüelles has long been a source of inspiration for many of those who look to the calendar and 2012 for guidance. He made waves with his book The Mayan Factor *and introduced us to what he calls the "Galactic Synchronization Beam." According to Argüelles, the Galactic Synchronization Beam is a time wave that is now triggering a new phase of galactic evolution. The following article is adapted from an interview with Argüelles conducted by Tami Simon in 1987 as part of the* Earthshift Series. *Here, Argüelles discusses his 30-year investigation into the advanced Mayan calendar, its origin, and how we may learn a lesson from it and forge a path beyond technology. He explains the Maya in the context of their "galactic viewpoint" and answers such questions as, What is a galactic beam, and how does it affect us? What other ancient calendar systems converge with the Mayan calendar and 2012? Were the Maya from another planet?*

UNDERSTANDING OUR PLANET THROUGH THE MAYA AND THEIR GALACTIC VIEWPOINT

Traditionally, we understand ourselves as being on this planet in this particular solar system. For instance, a system like astrology looks at the influences that affect us on this planet with relationship to the other planets in our solar system. We usually give very little consideration to the fact that there are other factors, galactic factors, that affect our solar system; that our solar system is actually a star system within a constellation of other star systems; and that we're operating within a larger field, the galactic field. This galactic field has many different energy beams that influence the different constellations. The different constellations themselves have effects on each other. The Maya understood their being on this planet in this galactic context.

Historically, we know the Maya as a civilization that existed in Central America and present-day Mexico and the Yucatán Peninsula. We know that their classic civilization, as it's usually referred to, had a duration of some 500 years, A.D. 400–830. When I talk about the Maya, I'm specifically talking about the classic civilization Maya.

There have been many Mayan art exhibits in all major U.S. cities. Many people have posed the questions, Who were the Maya? Why did their civilization disappear so quickly around A.D. 830? What were they doing with the most refined calendar that we have known on this planet, when it seemed that they were relative Stone Age people? They had no beasts of burden. They had no metallurgy. They did not use the wheel. In any case, who were these people who came up with this calendar and an amazing mathematical system? For most people, the Maya pose a great enigma. The archaeologists' answers are really not sufficient, and the mystery continues. In fact, we could say that the mystery of the Maya is one of the greatest mysteries of this planet.

As for me, I have studied the Maya since I was about 14, when my imagination was first tweaked by all of this. I plodded through some 30 years of working with this material until about two years ago, when a Maya by the name of Hunbatz Men contacted me, and I invited him to Boulder, Colorado, so we might meet. While in Boulder, he gave a talk on Mayan astrology. His information was what I needed to put the pieces of the puzzle together. The one juicy piece of information that Hunbatz Men left me with was that our star system represents the seventh such star system that the Maya have mapped, charted, and navigated. This led me to think, "Well, if that's the case, let's assume that the Maya were not from this planet originally, at least the seed Maya." From this position, I went back and reexamined the information and data I had. It all fell into place, and it all made sense. The Maya are what I refer to as "galactic surfers." Some of them were especially amazing, and I refer to them as "galactic masters."

THE MAYAN PURPOSE AND THE GALACTIC BEAM

The purpose of the Maya coming to this planet was very specific: to leave behind a definite set of clues and information about the nature and purpose of our planet at this particular time in the solar system and in the galactic field. The Maya came here at a very specific time and made their observations. They left behind observations about the relationship of our planet to at least seven inner planets and information about the nature of the galactic beam our planet is now passing through, and which it has been passing through since 3113 B.C. Once they made their calibrations about the relationship of our planet to this galactic beam and to the other planets and to the sun, their job was done.

One of the mysteries of the Maya has to do with their calendar, or what has been interpreted as their calendar. If the classic Maya didn't begin building their temples and cities until A.D. 100–300, why were they using a precise calendar with an initiation point equivalent to August 13, 3113 B.C.? This calendar is usually referred to as the "Great Cycle," and it runs from 3113 B.C. to A.D. 2012. The question is, Why did they have a calendar that begins in 3113 B.C., when they did not seem to have really begun to flourish as a civilization until the early centuries of the present millennium?

When I assumed perhaps the Maya were originally from outside of the solar system and from another portion of the galaxy, I also looked at the actual mathematical system that was being used to define their calendar. It was often assumed that the Maya invented the calendar to keep track of agriculture and planting seasons. I went back and assumed that the mathematical system existed first and was then adapted to a terrestrial calendar. As I looked at the mathematics, I was struck by the amazing quality of harmonic numbers in the calendar. The calendar was actually measuring some type of harmonic wave, some type of large wave or beam that had particular cycles of fluctuation and amazing harmonic properties. It quickly struck me that the Great Cycle isn't so much a measurement of time as it is the measurement of this planet's passage through a beam that is 5,125 years wide, or 5,125 years in diameter. Naturally, it's difficult to be aware that we are passing through a beam that is being generated from the galactic core and mediated through the sun, the local star. The Earth is passing through it, as are all the other planets, but it seems to have particular meaning for the Earth.

Every galaxy has a center. Many galaxies that we physically perceive are spiral galaxies with center points. The galactic core is an unimaginably dense central core of the galaxy that is the originator of galactic beams and the information. What is generated from the galaxies is very closely related to quasars and other such phenomena. We know that there is a continuous transmission of different types of radio waves, what we might call "information beams," being emanated and generated from the galactic core. The questions are, What is the nature of these beams? Why are they being transmitted? Is the information from these beams the information that generates life? Is the DNA code part of the information in these beams? How do these beams actually affect different planetary systems?

THE MAYAN SCIENCE

The Mayan science is based on principles of resonance. Our science, as originated in the seventeenth century, is based on matter—that matter is the ultimate reality and that it's the ultimate knowable reality. Modern science, however, has come to places where this is no longer clear. Modern science has gotten much closer to where Mayan science begins. Mayan science assumes that the key factors in universal operations are factors of resonance—vibratory cycles, or vibratory waves. These waves reach certain condensation points and become matter—atoms, subatomic particles, and so on. But the underlying nature of reality is vibration, resonance.

Resonance is a quality. The word *resonance* means "to sound again, to reverberate." Some type of medium allows vibrational signals to pass from a transmitter to a receiver. The vibrational signal is a frequency of some kind. In the Mayan perspective, all of reality is constructed of different levels of frequency.

One other key difference between Mayan science and modern science is that modern science generally assumes that the only knowable reality is the world of matter that we experience through our senses; it does not admit that there could be other realities, other dimensions coexisting with this reality. The properties of resonance that allow for the beginning understanding of other coexisting dimensions lie in the concept of the overtone, for instance. Every tone has a vibratory quality as well as overtones. The overtones occur in many different octaves, although in our reality we only hear one tone. This reality that we experience right now as we're sitting right here and listening or talking is one tone, but at the same time, there exists within it different overtones.. These different overtones are keys to different coexisting dimensions—the different dimensions of reality that coexist with the one that we usually assume to be the only real one.

APPLYING RESONANCE TO THE PATH
BEYOND TECHNOLOGY

Everything is in resonance. Let's begin with that principle. It really does take a radical shift to accept the implications of this principle completely. If everything is in resonance, ultimately, everything is in some kind of condition of harmony. There are different shifts, different waves, and particular cycles that dissolve and others that begin, but basically, there's a harmonic structure to all reality.

If everything is in resonance, what does that mean to us? It means that we are in resonance with the environment. It means that we are in resonance with the planet. It means that we are in resonance with the sun. As we know, the activity of our brain is measurable in different brain wave cycles and different frequencies. For instance, we know that the alpha frequency of our brain waves corresponds to the basic resonance of the planet, 7.5 hertz, megacycles—a very interesting illustration of resonance.

This opens the doors to possibilities that our present state of mind, our present science, our attachment to it, and the states of mind that are created by it do not allow. Through matching our own resonances and frequencies with key environmental resonances and frequencies, we can begin to see how we could draw energy from the environment directly—directly through the medium of our own bodies. This is a radical thought.

If we look around in nature, it's very interesting to note that we're the only creatures who use outhouses. We're the only creatures with insulation, clothes, and so on. All the other creatures are able to coexist with nature; they're able to moderate their body temperatures with the environment, and so on. It could be that we have the same ability, but we're not asking the right questions, or even worse, we've accepted the wrong answers and are not utilizing the full resonant receptive capacity of our being to establish the proper kind of environmental harmony.

THE EARTH'S RELATIONSHIP TO THE GALAXY
AS ENCODED IN THE MAYAN CALENDAR

I've been studying the Mayan calendar for some 30-odd years. I've also studied some other systems, most notably, the I Ching. I've always felt that there must be something in the Mayan calendar analogous or equal to the I Ching. As I've studied the calendar, it has become more and more clear, particularly in the last few years, that it had many other uses and purposes than just a calendar. It occurred to me that what we call the calendar is just one of the uses of an underlying code. We can look at the calendar, and the code is there.

It's important to understand certain principles related to holography, or what is called fractal mathematics. One is the idea that the whole is contained in the part, so that any part has the whole encoded in it. I began to see that the calendar was derived from the code. That's the most important point. Then I asked, What is this code? What is its meaning? I began to see that it had many other applications. I saw that the calendar, which is usually described as a continuously repeating 260-day calendar, was the code for a 260-unit matrix. When I looked at the Great Cycle, I saw that it, too, corresponded completely to the code. The Great Cycle consists not of 260 days, but of 260 katuns or 20-year periods, so it was the same code. From there, I began to see that the code had in it mathematical properties that related to virtually all other aspects of existence. In our present-day science, we have the periodic table of the elements. I began to see that the code upon which the Mayan calendar is based, and upon which the Great Cycle is based, is something like the periodic table of elements, only it's a periodic table of galactic frequencies.

Going back to resonance, the idea is that, in actuality, form follows frequency. First, there is a vibrational frequency that runs in set cycles, and from that, whatever forms exist are created. The frequencies have mathematical properties that can be translated into geometry. Geometry translates into form, and so the underlying code is actually the periodic table of galactic frequencies that govern the manifestation of all phenomena at many different dimensions. It's the universal code. It's a phenomenal, mind-boggling code.

OUR CURRENT SITUATION

According to the Mayan calendar, we're a little more than 25 years away from departing from the beam we've been in since 3113 B.C. When we look at the calendar as the representation of the beam, we see that this beam is divided into 13 subcycles, called baktuns, and that the present baktun, which is the thirteenth subcycle of this beam, began in 1618 and runs to 2012. Each of these subcycles is about 394 years long, so we're close to the end. What does this mean? It means that we are now at a highly critical moment. From 3113 B.C. to the present, this beam has been like a wave that has been building; going up and down, gathering force, and getting ever higher, stronger, more cumulative. While this planet has been passing through the beam, we have been extruding, through the medium of humans, the technological civilization that we see around us. In the last 200 years, the process of extrusion has really picked up. We've covered the world with our technology and with what I refer to as "our planetary exo-nervous system." We're at a point where we've reached saturation. In terms of the beam, we're almost at the point that can be interpreted as the saturation of technology in the world.

If we look at the beam as a wave, we see that the wave that has been building up for 5,100 years is about to break. The precise breaking point of the wave occurred August 16 and 17, 1987, the Harmonic Convergence. Imagine it like this: You're standing on the beach. You watch a wave wash in from the sea. The wave crests, and then it breaks, and then it just washes in. When the wave hits the beach in this metaphor, it will be the year 2012. When the wave hits the beach, in 2012, we will be prepared to enter into galactic civilization or interplanetary civilization. This last 25-year period is a very interesting, very significant, very critical time for the planet. This is an extremely critical moment from the galactic point of view. We could say that the higher intelligences are waiting for us to become exhausted with our technology and to realize that, ultimately, our technology can't help us.

One of the most radical aspects about my book *The Mayan Factor* is that it posits the possibility that our technology is really no longer necessary. The evolved species known as Homo sapiens has been affected by this beam. One of the ways it's been affected by it is by extruding technology, extruding an artificial environment, and extruding artificial means of communication. This obviously is an evolutionary phase, as we might call it, an evolutionary transition, but we also have to look at the fact that from the point of view of the planet, we're on the surface. The planet is an evolving, living organism, and we are the most active beings on its surface. In a positive sense, we could say, "The purpose of extruding this technology has been to get us to the point of becoming really self-conscious and seeing ourselves as planetary beings, as planetary organisms. Perhaps the only way we could have come to this point is through developing this technology and this type of communication." This communication, as we know, has shrunk the world into an instantaneous communication system.

If you look at this from a total planetary point of view, we have entered a very dangerous moment because, in the last several hundred years, we have created a civilization that is reliant on fossil fuels and different types of metals, precious minerals, bare metals, and so on, that are taken out of the Earth. What has happened in our technological episode, on this spaceship we call Earth, is that we have virtually used up all of our spaceship's energy stores. Even if we say we've got oil to last another 40 years, that's not very farsighted. What happens after that? The issue really is, "We shouldn't use up any more of the energy stores." Certain energy stores accumulate, and when you need them, you can draw on them. But you don't deplete them in one big technological binge. This is exactly what we have done. The path beyond technology is an absolutely necessary path for us to take. During the time that we have been depleting the energy stores of spaceship Earth, we have also succumbed to a one-dimensional view of reality. In this one-dimensional view of reality, the circuits of our belief systems have been turned upside down. The mainstream, mass mind-set believes that the use of fossil fuels and chemicals, which produce incredible pollution, is superior to the use of solar energy. We *believe* that. In the technological binge, we've become completely intoxicated, one-dimensional, or amnesiac. We believe that the dark is better than the light.

2012: TIME TO WAKE UP

We must wake up. Right now, we live in a world where human consciousness basically is a function of what we call "mass mind." The mass mind is controlled and manipulated by—and believes in—governments and the whole range of institutions that are the key factors in modern life. Governments, institutions, and the mass media continuously put out information of a negative nature, negative energy, so that mass mind thrives on negative energy. Governments and institutions support this creation of negativity, and so humanity as a whole is in the grip of a very bad type of cultural hypnosis or civilizational trance. What matters is that we wake up so that we can enjoy the transiting of the beam. We have created such an imbalance through our technological binge, depletion of resources, and pollution of the planet that we must wake up if we want to continue enjoying the planet. Our waking up will mean that we actually see that technology can't help us and that we have to go beyond it. If we don't wake up, the planet will naturally self-destruct because it is already so far out of balance.

We're precisely at the point in time where the difference between a considerable number of us waking up and taking command and not waking up is the difference between the planet going into self-destruction and the planet continuing on its evolutionary path. To continue with the image of spaceship Earth, although we have used up the reserve energy supplies, we still have the central source, the sun, and our own very sophisticated bio-psychic instrumentation, our very beings, our human bodies. In terms of spaceship Earth, the wrong crew is in command, and it's time for a mutiny.

PARALLELS AMONG THE MAYAN CALENDAR, THE I CHING, AND THE BOOK OF REVELATIONS

There are actually many points of convergence among the Mayan galactic code, the I Ching, the Book of Revelations, and many other systems. The reason for this is very easy to explain. We're all operating in the same field. Be it ancient China, ancient Egypt, ancient Israel, or ancient Mexico, the human receivers picked up the same information beams, translating them into particular languages conditioned by already-developed beliefs in those particular areas.

In one case, you have the development of the I Ching in China with its mathematical code, which is exactly the same as the mathematical code of DNA and exactly the same as the mathematical code of the harmonic number progression of the Mayan code, a binary progression. Or you have the mathematical code of the Book of Revelations, which is based on 7 and 13, and 144 and 144,000; it's the same number code as the Mayan code.

The Mayan code is the master code. It's the galactic code. It's the master program. The other codes are there, as I said, because humans everywhere can pick up the same information from the same beams that are affecting the planet. Most of these codes and systems, like the Book of Revelations and the I Ching, were developed several thousand years ago. In these earlier phases before technology became the dominant factor, human receivers tended to be more operative. Right now, our receivers are quite shut down. There are people awake, here and there, making efforts to get other people to awaken. From the point of view of the code, being awake means opening up our receivers. This can only be done by having a genuine change of heart; a change of heart about what is happening in our lives, about what is happening on our planet, and realizing that if we are to continue, we must open up our receivers so that we can begin to receive the information and know what to do next. One might ask, If this is all laid out, what can we do? The fact is, it is all laid out only in some regards. We are so asleep to what *is* laid out for us that we can't take advantage of it.

The Maya made a visitation. They made a very skillful intervention into the biosphere of this planet and left a civilization behind that looks like a lot of other civilizations. It's very carefully disguised. Through my study of the calendar and my study of other code systems and calendars on the planet, I saw that all of them were contained in the Mayan code—particularly, the DNA code and the I Ching code, which is contained as the central component, which I show in *The Mayan Factor*. There are no other codes like this on the planet.

GALACTIC SYNCHRONIZATION AND UFOS

In 2012, we will be at a place on the spiral, where only maybe one turn up on this particular beam, where we were in 3113 B.C., and what that means is that there's a fresh start, and I think it means for ourselves here that there's an evolutionary jump. The cycle that I've been talking about, the Great Cycle, is only a subfactor of another larger cycle and all cycles—they go up (macro), and they go down (micro). The larger cycle that we're dealing with is an approximately 26,000-year cycle that also ends in 2012. If you go back 26,000 years, this cycle roughly encompasses the evolutionary stage of Homo sapiens. The peak of the last Ice Age was in 24,000 B.C. This is when what we call "modern humanity" emerged. So, this 26,000-year cycle actually encompasses five Great Cycles. We're in the fifth and last of a set of Great Cycles that began in 24,000 B.C. It's on this basis that we're looking at 2012 as being a very, very critical juncture. It's what I refer to as "galactic synchronization," and it also is a major evolutionary shift. The next evolutionary shift will occur in 2012.

There are other dimensions of reality, and there are more evolved stages of being and intelligence than ours. The universe is benign and compassionate. No one's out to destroy us. UFO activity has increased since we entered a critical pollution stage, which followed the introduction of radioactive waste into the environment in 1945. It's been since that point that the modern phenomenon of UFOs has been with us. There is definitely a relationship between the Mayan factor and the UFO phenomenon, and I think that we can anticipate that the UFO activity will intensify. We must understand and realize that we are dealing with galactic civilization. That realization will facilitate a return to our own highest being, to our own receivers opening up, and the information, which right now we refer to as the information of the Mayan factor, will be information easily accessible to everybody.

The Nine Underworlds
EXPANDING LEVELS OF CONSCIOUSNESS

CARL JOHAN CALLEMAN, PH.D.

*Swedish researcher Carl Johan Calleman claims that there
is a Mayan prophecy that "the world of consciousness will be
born" during the June 5–6, 2012, Venus transit. Ancient Maya
believed that there were nine Underworlds, each representing a
different level of consciousness and a different segment of time
according to their calendar. According to Calleman, we are
currently in the Eighth Underworld. He says, "This eighth level
may be referred to as the galactic frame of consciousness, as it
will step-by-step lead humanity to identify primarily with the
galaxy. The highest level of consciousness, the Universal, will
be attained through the workings of the Ninth Underworld in
the year 2011 and will result in a timeless cosmic consciousness,
and a citizenship in the universe, on the part of humanity." In
the following excerpt from his book* The Mayan Calendar and
the Transformation of Consciousness, *Calleman explores
the Nine Underworlds and the concept of the last level, the
Universal Underworld. He answers such questions as, What
will happen when we reach the last Underworld in 2012,
according to Mayan prophecies? What type of consciousness
will we encounter in the Universal Underworld and 2012?*

In Mesoamerican mythology, there were Nine Underworlds. Although it is not possible to gain a detailed understanding of the origins of these Underworlds, we may surmise that they are nine sequentially activated frames of consciousness mediated by the Earth's inner core.

Today, we know that the world did not begin only 5,000 years ago with the emergence of the first written language, the first pyramids, and the first nations centered on pharaohs. Modern science has shown beyond a doubt that the present universe came into existence much earlier than when the Great Cycle started. It was born some 15 billion years ago as matter was first created from light in the so-called Big Bang. Since then, our galaxy, solar system, and planet with its biological organisms have all come into existence. Hence the Great Cycle did not begin from nothing. It had considerable previous evolution to stand on.

How does the Mayan calendar fit with this great age of the universe? Interestingly, the Maya were aware that the world was much older than 5,125 years. On a stela discovered at the ancient site of Coba on the Yucatán Peninsula, the creation date of the Long Count is placed in the context of a hierarchy of creation cycles of 13 3 20^n (13 times 20 multiplied n times by itself) tuns. Thus there is a whole series of creation cycles like the Great Cycle (which itself is 13 3 20^2 tuns). Creation may then be regarded as a composite of several creations, each being built on top of another to form a pyramidal structure, and of these cycles the Great Cycle is but one. Today we can still see this idea of several different creations in ceremonies of the Maya, where, for instance, monkeys from a "previous creation" are sometimes enacted. Each of the Nine Underworlds of Mesoamerican mythology is a different "creation" generated by a cycle 20 times shorter than the one upon which it was

UNDERWORLD	SPIRITUAL COSMIC TIME	PHYSICAL EARTH TIME	INITIATING PHENOMENA	SCIENTIFIC DATING OF INITIATING PHENOMENA
Universal	13 x 20 kins	260 days	?	
Galactic	13 x 20^0 tuns	4,680 days (12.8 years)	?	
Planetary	13 x 20^1 tuns	256 years	Industrialism	1769 C.E.
National	13 x 20^2 tuns	5,125 years	Written language	3100 B.C.E.
Regional	13 x 20^3 tuns	102,000 years	Spoken language	100,000 B.C.E.
Tribal	13 x 20^4 tuns	2 million years	First humans	2 million years
Familial	13 x 20^5 tuns	41 million years	First primates	40 million years
Mammalian	13 x 20^7 tuns	16.4 billion years	Matter; "Big Bang"	15–16 billion years

FIGURE 14: The durations of the Nine Underworlds in both spiritual and physical time, and some of their initiating phenomena.

built. This is why the most important of the Mayan pyramids—the Temple of the Inscriptions in Palenque, the Pyramid of the Jaguar in Tikal, and the Pyramid of Kukulcan in Chichén Itzá—were all built as hierarchical structures with nine levels.

Although these Underworlds have sometimes been translated as "hells," I am convinced that this is misleading. It does not make sense that the Maya had built their nine-story pyramids in honor of some "hells." Underworlds are related to the sequentially activated crystalline structures in the Earth's inner core. (The origin of the Christian hell was the Norse Earth Mother, Hel, who ruled the Underworld and was portrayed as a fearsome force in the patriarchal Great Cycle.) The beginning dates of the creation cycles generating these Nine Underworlds and some concomitant events (seeds)—which, in accordance with modern scientific dating, occur at the beginnings of the Underworlds—are summarized in Figure 14.

The Great Cycle, creating the Sixth Underworld and generating a national frame of consciousness to which we have until now given all our attention, is thus just one among several such creations, or Underworlds. This particular Underworld was built on the foundation of five lower Underworlds that had generated more limited frames of consciousness. The nine-story Mayan pyramids are thus telling us that consciousness is created in a hierarchical way and that each Underworld stands on the foundation of another.

Before continuing, we should summarize the various time periods that are part of the tun-based (360-day) calendrical system of the Maya. Each of the nine major creation cycles is based on a period that is a multiple of the tun. The tun is multiplied by 20 different numbers of times and in such a way that a hierarchical system of time periods is generated. (Note the exception, however, that there are 18, rather than 20, uinals in a tun).

Each of the Nine Underworlds is developed through a sequence of Thirteen Heavens with seven days and six nights, beginning with 13 *hablatuns* in the Cellular Underworld and continuing upward through *alautun, kinchiltun*, and so on. Hence, at every new level, the Thirteen Heavens (of a 20-times-shorter duration than the level below) serve to develop a higher frame of consciousness. Thus, each Underworld is associated with a certain frame of consciousness: cellular, mammalian, and so forth. Each Underworld is also associated with a certain frequency of creation, the frequency with which the energy changes take place. At all levels, the energy changes are the result of alternations between seven days and six nights. As an example, the sixth level (the Great Cycle, or National Underworld), ruled by a commonly mentioned deity by the name of Six-Sky-Lord, has a baktun period (394 years) of energy changes that serve to develop

THE PLANETARY UNDERWORLD

1755 C.E.  2011 C.E.

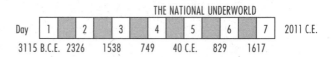

THE NATIONAL UNDERWORLD

Day	1	2	3	4	5	6	7	2011 C.E.

3115 B.C.E. 2326   1538   749   40 C.E.   829   1617

THE REGIONAL UNDERWORLD

(100 B.C.E.–2011 C.E.)

5869 B.C.E.

	Day 7 of the Regional Underworld	2011 C.E.

FIGURE 15: Relationships among the Regional, National, and Planetary Underworlds presented in an accurate timescale and illustrating the increase of the development from seed to fruition. On this scale, the lowest Underworld (the Cellular) would measure some 25 miles.

its national frame of consciousness. In Figure 15, the different energy changes of the Regional, National, and Planetary Underworlds, each taking place with a certain frequency, are shown in a correct comparative scale. We see how the National Underworld began a given time after the seventh day that the Regional Underworld was established, and then in turn how the entire Planetary Underworld falls within the seventh day of the National Underworld. In the scale used in Figure 15, the longest creation cycle, the Cellular Underworld, would stretch some 25 miles (40 kilometers). This is a good illustration of the phenomenon that evolution is accelerating.

As of this writing (2003), we have come to the third day of the creation of the Eighth Underworld, the Galactic, which in turn stands on the Planetary Underworld. There are now only two more Underworlds to go—the Galactic (12.8 years) and the Universal (260 days)—until we reach the highest level of creation. In this hierarchical structure, the different frames of consciousness (Regional, National, Planetary, etc.) do not replace each other, nor do they follow one another. Instead they add to each other, so that the creation of all the Underworlds, and thus the climb of the nine-story cosmic pyramid, will be completed at the same time: October 28, 2011, in the Gregorian calendar. As we will shortly see, each Underworld is also associated with a specific yin/yang polarity.

Creation thus brings about the evolution of the cosmos through nine distinct Underworlds, going from the lower Underworlds, where the evolution of consciousness is manifested in physical ways—such as galactic matter and biological species—to the increasingly more ethereal or spiritual expressions of the higher Underworlds. Looking at the ascent of the cosmic pyramid from the perspective of modern knowledge, we may identify a sequence of different frames of consciousness, each associated with a particular Underworld. Because these frames expand from the Cellular to the Universal, the purpose of creation seems with every Underworld to be to expand and elevate this frame, ultimately making possible the advent of the Universal Human.

THE UNIVERSAL UNDERWORLD . . .

Then I saw a great white throne and Him who was seated on it.
—Revelation 20:11

Again, what will make you realize what the day of Judgment is?
The day on which no soul shall control anything for (another) soul
and the control shall be entirely Allah's.
—Cleaving Asunder Surah 82:17–19, the Qur'an

The spiritual beings will remain to create one world
and one nation under one power, that of the Creator.

—Hopi prophecy

The Universal Underworld in 2011 is what all of creation has been waiting for: its very purpose. It is when all things are brought together and all the conflicting ways of being, acting, and thinking will be resolved and unified in a light that makes it possible for everyone to understand everyone else and every-thing at once. All limiting thoughts will disappear. The Ninth Underworld may be seen as a gift from God, since it is not only about creating balance, but also about the enlightenment given to humankind as an expression of divine grace. This is when we will fully understand why the cosmic plan was designed the way it was, and we will overflow with gratitude to the Creator. At the same time, we will recognize our own divinity, for there will be no separation between the divinity of the Creator and our own.

Not surprisingly, there are several myths from different parts of the world that talk about Nine Worlds. In the Norse tradition, the cosmos was believed to be made up of Nine Worlds, and to the Hopi Indians, there are also Nine Worlds. In the unification of science and myth that is now taking place, these and a number of others can be recognized as referring to the Nine Underworlds of the Mesoamerican tradition. What makes the Mayan calendar tradition unique is its precision.

Paradoxically, as the Mayan calendar system draws to a close, time will be experienced on one hand as moving faster than ever before, and on the other hand as if it were not moving at all. The enlightened consciousness developed in the Universal Underworld will be pulsed onto humankind in a wave move-ment of the Thirteen Heavens that covers a period of only

260 days (or possibly, if it is one-twentieth of the Galactic Underworld, 234 days; I cannot tell yet). This reflects a frequency of change of the heavenly energies that by far surpasses anything anyone has ever experienced. Yet there is every reason to believe that, at the same point, "time" will come to an end, since "time" is an experience that is predominantly mediated by the left-brain hemisphere. In actuality, the experience of time exists only in a world dominated by the imbalance created by duality. As balance between the two hemispheres is created, instead of time we may expect to experience pure being moment by moment.

This paradox, of course, is not easily resolved by someone whose mind is still dominated by the dualities generated by lower Underworlds. The ego that was established to serve the dualist mind in navigating their changing waters and to maintain a sense of continuity for the individual in the midst of them simply cannot survive at the high frequency of the Universal Underworld. The ego is not consistent with the unitary field of light that will then be ruling the world. In a dualist frame of consciousness, the ego was an important tool for survival; in a unitary frame of consciousness, it will jeopardize the survival of the individual. In an Underworld dominated by unity and enlightenment, acting and thinking based on a dualist frame of mind will become impossible. At the frequency of change that will prevail in the Universal Underworld, with shifts between days and nights taking place every 20 kin, the mind has to be disengaged if it is not to lead the individual to complete heartbreak or personal collapse. Even with proper preparation, for a person who is well aware of the workings of the cosmic plan, the conflict between the 5,000-year-old dualist mind and the unitary consciousness of the Universal Underworld can only be successfully resolved in one way—through the excision or "slaying" of the ego and the ensuing creation of space for the enlightened way of being to take over. This is what Kalki proposes to help us do. Such external help may be sought and welcomed by some, but others may come to attain the enlightened state simply by surrendering to divine grace. The intent and humility of each one of us will determine the outcome.

As we will then be endowed not only with a galactic but also with a nondualist, universal frame of consciousness, these frames will also make it possible for us to see the events of the past from a different perspective. Being fully immersed in the present, without the dualist mind's need for maintaining continuity with the past, the Universal Underworld will allow for true forgiveness. The Book of Revelation refers to this in verse 21:4: "And God shall wipe away all tears from their eyes; and there shall be no more death, neither sorrow, nor crying, neither shall there be any more pain; for the former things are passed away." From the perspective of the Universal Underworld, the former things will have passed away.

By 2011, the dominance of the dualist mind will wither, and all the conflicts of humanity originating in lower levels of consciousness will be resolved. From the perspective of the enlightened state, the old order will no longer be real. The advanced technologies developed by the dualist mind will find the place they were always meant to have, in service to humanity and the living cosmos, not as tools of domination. By then, not only the old monarchic rule, but also democracy, will be things of the past. (If everyone lives in unity and harmony with the Divine, why elect someone to rule them?) All hierarchies will have crumbled. With the end of duality, the dominance of one soul over any other will naturally come to an end, and so there will be no need for government (which is like the ego of the nation) to steer people through conflicting interests. All human beings will, in a much deeper sense than at present, be recognized as having equal value—each as her or his own manifestation of the Divine. In the process of climbing to the universal level of consciousness, all limiting thoughts will disappear.

Once this has happened, we will be able to fully experience the unity with All That Is, moment by moment, or frame by frame as it were. Such an experience, frame by frame of consciousness, will allow for total sensual enjoyment in the present moment. Lacking the mind's need to maintain continuity, we will experience complete freedom, and our status as puppets of the divine process of creation will come to an end. There will no longer be an experience of separation between human beings and God. If we do not experience ourselves as gods, we will, at the very least, experience ourselves fully as the manifestations of the Divine that we truly are.

As a corollary, to wait around to see what will happen in the year 2012 is to miss the point totally. It will simply not be possible not to be enlightened after October 28, 2011, or at least not from a certain time afterward when the new reality has definitely manifested. With a dualist mind, it will be impossible to be in resonance with the new unitary divine reality under one power, the Creator. It would thus seem wise for all of us to prepare ourselves, beginning today, by immersing ourselves in the cosmic flow of time and, in all possible ways, seeking to transcend the influence of dualist Underworlds on our thinking, acting, and being. After all, the Universal Underworld will favor an enlightened state of being of love and joy, and once this has been established, no reversal to duality will be possible.

This will mean the return of an apprehension of the living cosmos. The enlightenment that this cycle brings will allow for true and complete healing and forgiveness of the past, which humans will also extend to God, whose relationship with human beings (because of the duality of lower Underworlds) has sometimes been perceived as antagonistic. When the energy of 13 Ahau is reached in all the Underworlds, no filters will block our vision of divine light or our communion with the Divine. We will come to live in the New Jerusalem, having reached enlightenment after climbing the Nine Underworlds and completing the 108 movements of Shiva. After the building of its ninth level has been completed, humankind will be ready to stand on top of the cosmic pyramid. Humanity will be complete, living fully in the present, at a higher level of awareness, and with a pure enjoyment of living. The creation cycles that have served to build the cosmic pyramid, partly by making human beings into puppets, will come to an end. The life of Universal Humankind will begin.

. . . AND BEYOND

As we have seen in the history of humankind, it takes a certain amount of time until a new frame of consciousness finds its full manifestation. It is only realistic to expect that something similar will be the case after the enlightened universal frame of consciousness is finally established as a Heaven on the day 13. 13. 13. 13. 13. 13. 13. 13. 13. 13 Ahau (October 28, 2011). Even if the status of human beings as puppets of creation disappears through the influence of this energy, it seems unlikely that this will immediately be expressed in stable forms. Hence, the year 2012 may also be a period during which many people will have to find ways of adapting to the new frame of cosmic consciousness. If nothing else, we will need to adapt to the fact that everyone around us is now enlightened and has full faith that a millennium of peace, the Golden Solar Age, has finally dawned on Earth. This does not mean that a "new cycle" will begin. It is the end of cycles.

With the ego slain in the Universal Underworld, the Heavens will favor a consciousness that is devoid of inner conflicts and that produces no external conflicts. The year 2012 in particular, however, will be required for this enlightened state to settle in a number of individuals who may not have been fully prepared for it. For the final settling of the enlightened consciousness of the Universal Underworld, the celebrations on the occasion of the Venus transit of June 6, 2012, will be very important. They will be expressions of gratitude to the divine cosmos for having brought us to where we are.

Both Mayan and Christian sources talk about an end to death at the end of time, and Eastern traditions talk about the enlightened state as deathless. The ancient traditions all point in the same direction: a timeless, enlightened state of cosmic consciousness. We may then rightly ask, Who in such a state needs a calendar? The Mayan calendar will have come to an end, and its use becomes meaningless. The calendar is like a staircase that is absolutely necessary to bring us to the top but serves no purpose after we reach it. In the year 2011, at a point when all Underworlds will have attained the tzolkin energy of 13 Ahau, the divine process of creation is completed. What the "future" will be is not predictable from the Mayan calendar, because the wave movement of all its Underworlds will have come to an end. It is the nature of the enlightened human being to be totally free to live according to his or her choosing. With such a freedom, and liberation from the cosmic plan, the further destiny of humankind is impossible to predict. We will be completely free to chart our own destiny. Humanity will live in true freedom, joy, and peace.

2012 and the Maya World

ROBERT K. SITLER

The number of 2012 experts and forecasts is growing by the day.
Many 2012 leaders claim to know what the Maya calendar truly
means, but within the movement, there are many opposing views.
In this essay, Robert Sitler examines the most well-known experts,
gurus, and leaders responsible for the widespread attention to
the Maya calendar and presents a counterview to some popular
predictions. From the indigenous Maya to those who tout Maya
wisdom with tenuous ties to Maya heritage, Sitler highlights
some of the influential 2012 teachers in the context of the modern
Maya culture. He asks, Is the swirling 2012 conjecture more a
modern phenomenon than an ancient prediction? How much of
the 2012 hubbub is actually rooted in Maya historical findings?

Hundreds of Internet Web sites in dozens of languages, a
rapidly growing assortment of books, videos, international
workshops, and electronic discussion groups speculate wildly
about the significance of the Maya calendar and the portentous year 2012. In general, this highly creative speculation
focuses expectations on an approaching period of radical planetary transformation that will include cataclysmic changes and
humanity's subsequent emergence into a new age of expanded
consciousness. Predictions concerning how this global shift

will take place include spectacular scenarios such as a sudden inversion of the Earth's magnetic poles, the descent of a spiritually uplifting galactic beam, and even the arrival of master spiritual beings from distant stars. Hyperbolic conjecture aside, the December 21, 2012, date is simply the last day of the current *pik* (pronounced "peek") cycle,[1] a period of 144,000 days in the ancient Maya Long Count calendar roughly equivalent to 394 years. More significantly, it marks the end of the thirteenth pik, the final period of a far larger calendar cycle measuring 1,872,000 days that began on August 11, 3114 B.C.E. and that will come to fruition more than 5,125 years later on the 2012 Northern Hemisphere winter solstice.[2] It is important to point out that even this so-called Great Cycle in the Long Count merely represents a minor component in far larger Maya chronological periods that theoretically extend backward and forward in time in a system of exponentially increasing temporal cycles without beginning or end point. The date is hardly, as some mistakenly claim, the "end" of the Maya calendar.

Although the 2012 date itself is indeed part of the Maya heritage, surprisingly, careful investigation of the major trends within the 2012 phenomenon reveals few substantive connections to the Maya world. Even the authentic links that do exist between Maya culture and 2012 ideology can get lost in a confusing patchwork of poorly documented competing New Age theories that are sometimes based on inaccurate facts and misinterpretation of Maya traditions. For example, in spite of the claims of some 2012 proponents concerning the centrality of December 21 in ancient Maya culture, the 13 pik date only appears on a single occasion in the entire known corpus of Maya hieroglyphic text.[3] Even in this tiny text, there is no mention of the events described in current popular speculation. Similarly, the varied and elaborate sets of "Maya prophecies" found online, the majority in Spanish, are largely New Age creations lacking substantive basis in Maya sources.

The relative absence of authentic Maya cultural material in the 2012 phenomenon should not be surprising because, apart from several notable exceptions, those most actively promoting the significance of the date are not Maya and have had few opportunities for significant practical interaction with contemporary Maya societies. These Westerners do not speak any of the 30 Maya languages and have minimal experiential familiarity with Maya traditions. While there are now probably more than 10 million Maya living today, thus far they have participated little in the 2012 phenomenon, as only a very small number have had any prior exposure to the topic.[4]

The Long Count calendar from which the 2012 date emerges fell into disuse well before Spanish invaders arrived in Central America in the sixteenth century.[5] Knowledge of its rediscovery by Western academics in the late nineteenth century has reached extremely few Maya living today. School curricula in the Maya world do not typically reflect native values and include little emphasis on the Maya cultural heritage. As a result, few Maya students have the opportunity to learn formally about their own culture except for the relative few at the university level. Those exceptional Maya who actually know of the 2012 date must have learned of it directly or indirectly through contact with non-Maya sources. While many Maya have long spoken of a coming time of great changes, the specific references to 2012 are a recent addition. Such reclaiming by the Maya of their own cultural heritage is yet another step to their eventual freedom from oppression by their Hispanic compatriots.

The few individual Maya who actively participate in the 2012 phenomenon tend to be heavily involved with foreign associates, primarily speak Spanish rather than a Maya language, and live a decidedly nontraditional lifestyle. Not surprisingly, their contributions often blend traditional Maya beliefs with an amalgam of New Age spirituality that includes visits from extraterrestrials, prophetic crystal skulls, and origin stories going back to the lost city of Atlantis. Confusing matters further, several non-Maya individuals pose as Maya elders and spread their own versions of 2012 lore. While many involved in the 2012 phenomenon seem well intentioned and promote noble aims, some Maya resent that a few also benefit from personal financial gain and notoriety within New Age spiritual circles.

Perhaps the best known of the real Maya teachers among New Age spiritual practitioners has been Hunbatz Men, a Yukatek Maya spiritual guide from the Itza lineage. At its core, Men's is a loving message presented with a poetic awareness of language. His valuable work as a Maya linguist draws attention to the multiple levels of meaning embedded in many Yukatek Maya words. Although a few admirers occasionally romanticize Men by presenting him as a traditional Maya community elder, his highly creative and multifaceted life has unfolded primarily in nontraditional urban centers. He studied gnostic philosophy and fine arts in Mexico City; worked as a commercial artist in New York City; and currently resides in Merida, the bustling capital city of the Mexican state of Yucatán, where he serves as a spiritual guide. His childhood did indeed unfold in a small Maya community under the spiritual tutelage of his uncle, don Beto, but his public work as a Maya spiritual teacher began in earnest in the 1980s when Men was already in his 40s. In 1985, he began working with Dr. José Argüelles, a man who deserves major credit for launching the 2012 movement as a social phenomenon in 1987 with the success of his book *The Mayan Factor: Path Beyond Technology*. Men initially shared many of Argüelles' artistically bold visionary approaches, including the adoption of a newly invented "Maya calendar" known as Dreamspell, which has become better known in the West than the real 260-day Maya calendar that has been in continuous use for millennia. More recently, however, Men has distanced himself from the work of his former associate. For example, Men no longer believes, as he once did, that the year 2012 has any extraordinary significance.[6] Men has been particularly active in visiting Maya archaeological sites in Mexico and Guatemala, leading group tours of New Age spiritual seekers to reconsecrate the sites as initiatic centers in preparation for a coming "Itza" era. He recently organized a group tour in the Maya world with the spiritual teacher Star Johnsen-Moser in order to donate one of the famous Maya crystal skulls to the sons of the late Maya elder Chan K'in (Viejo), representatives of the Lacandon Maya. The crystal skulls are a phenomenon unto themselves—13 quartz skulls known for their capacity to transmit information through "cosmic rays." According to Men's publications, since ancient times, the Maya have lived in several parts of the world, including Egypt, India, and even Atlantis.[7] Even though Men's teachings and spiritual work have inspired many spiritual seekers, most Maya cultural traditionalists and academic literature on the subject do not corroborate his beliefs concerning extraterrestrial masters, worldwide Maya influences, and Atlantian origins—all ideas more similar to those of Edgar Cayce than those of the Maya. Still, his genuinely warm demeanor, transparently positive intentions, and, most important, the fact that he actually is Maya establish sufficient authority for him to have attracted many sincere and dedicated students. Because he is Maya, Men's participation in New Age spiritual settings establishes an air of indigenous authenticity to teachings that few are in a position to question.

A man born in Mexico City called Quetza Sha is an associate of Hunbatz Men and refers to himself as an Aztec-Maya shaman. He directs what he calls a "Galactic University," which offers courses on karma, transpersonal psychology, and sacred geometry.[8] Sha's beautifully designed Web site says that he is a master from an esoteric group called the Sovereign Solar Order of Chichen-Itza. He says that the year 2012 will bring about a positively transformed humanity, a unified "solar tribe," and the birth of the age of the Aztec sixth sun. He states that the Maya "came from the stars," "from Pluto, Neptune, and Uranus," and then "disappeared, without any apparent reason, hundreds of years ago,"[9] a statement that leaves one wondering if millions of contemporary Maya do not share the genetic heritage of their ancient ancestors. While Sha's teachings offer uplifting direction for his students, his message, like that of many others in the 2012 movement, seems largely tangential to the realities of the Maya world.

Dan Rios is a New Age spiritual teacher who works under the name Tze'ec Ba'lam.[10] Like Sha, he is also of partial Maya ancestry and uses the honorific title Alom Ahaw. Ba'lam explains on his Web site that "the last recorded date on the Tzol'kin [sic] Sacred Mayan calendar is December 21, 2012, and the great cleansing of Mother Earth is occurring even now."[11] Given the perpetual nature of the ritual 260-day *cholq'ih* calendar, which has been in continuous use for thousands of years in the Maya highlands, the idea that it has a "last recorded date" is hard to understand.

K'iche' Maya elder Alejandro Cirilo Pérez Oxlaj stands out as the most traditional and respected of the few actual Maya teachers connected with the 2012 phenomenon. Don Alejandro, affectionately known as *tata* (father), was born in Quetzaltenango, the second-largest city of Guatemala, and was raised in the nearby village of San Francisco del Alto. According to his wife, he received his sacred *vara*, the bundle of holy *tz'ite* seeds cherished by highland Maya oracles, from his father at the age of 13. In doing so, don Alejandro's father established the thirteenth generation of Maya spiritual guides in his family. Later in life, in a revelatory experience, "invisible beings" gave don Alejandro the name Wakatel Utiw (Wandering Wolf)[12] and a mandate of being "the Voice of the Jungle" and a "messenger of the Maya."[13] From his current home in Antigua, Guatemala, he heads the Consejo Nacional de Ancianos Mayas de Guatemala, perhaps the largest of several multiethnic associations of Maya elders. As part of his mission, he travels the world as a representative of the Maya, sharing prophecies and a message of loving respect for all beings in preparation for the coming period of world transition. Most Maya *ajq'ijab'*,[14] or spiritual guides, who know don Alejandro's work admire him greatly—his humble nature and generosity of spirit in particular.

Even in the case of this revered Maya elder, however, a few elements of his teachings mirror the extraordinary work of Hunbatz Men and José Argüelles that originally shaped the 2012 phenomenon. In one recorded talk, for example, he refers to repeat visitations by Pleiadian masters to the Maya world and elsewhere on the planet.[15] This description of mythical masters from afar parallels an important episode in the *Popol Vuh*, the sacred K'iche' Maya holy book: the first four humans, the supernatural ancestors of the K'iche', travel to and from a great city in the east bringing benefits for their descendants. In what appears to be a revision of this ancient tale, the holy city the east has become a constellation of stars renowned in New Age circles. Online sources quote don Alejandro as purportedly stating,

Crystal skulls were brought from the Pleiades, they speak of
reincarnation. . . . The Mother of Mayan science is Pleiadian.
These skulls must go home where they belong. This is the science of
the future after 2000. Mayan actually means "my people."
They first came to Atlantis. Our first pyramids are there,
underwater. In the Bermuda area. Tulan was the name of the
city of Atlantis. . . . There have been many visitations from the
Pleiades, then they return there, then they come back here again
and come down to Asia, Tibet, Kaimaya, and in India also.
"Nagamaya" they were called. There are glyphs that speak of these
visitations. . . . In 2012, these wise ones will return again.[16]

The references in this passage—crystal skulls from the
Pleiades constellation, the existence of Atlantis, an ancient
Maya presence in Asia—clearly recall the teachings of Hunbatz
Men and José Argüelles. None has a basis in known historical
Maya sources, and none has the backing of any serious academic
investigator, Maya or non-Maya. Because such online quotes of
don Alejandro have questionable references and may be inaccurate or poorly translated, they might not accurately represent
his beliefs. However, because don Alejandro conducted tours
of the Maya world alongside Men, who was, at that time, an
associate of Argüelles, it would not be surprising if they all
shared some common beliefs. One alleged don Alejandro
quote states, for example, that a time of radical changes "shall
be fulfilled in Twelve Baktun and it began on August 17, 1987,
the Harmonic Convergence. . . . The Harmonic Convergence
began the Twelve Baktun, Thirteen Ahaw and will end in
2012 or 2013 by your calendar."[17] The Harmonic Convergence
referred to here originated in the 1971 work of Tony Shearer
and was popularized by José Argüelles. In retrospect, the date
marked a critical point in the growth of the 2012 phenomenon.

Unfortunately, the Harmonic Convergence date has no basis at all in Maya culture and is only loosely based on one person's interpretation of Toltec lore. The attribution of these statements to don Alejandro in no way diminishes his status as a spiritual teacher or Maya leader. However, it does demonstrate the ease with which non-Maya ideas can become confused with the teachings of even the most traditional sources.

While little content in the 2012 phenomenon has substantial basis in Maya culture, there are at least two legitimate historical sources and several minor social currents within the Maya world that do actually refer to epochal change and planetary renewal, the core themes of the 2012 phenomenon. The older of the two historical sources is a little-known ancient hieroglyphic text found on Monument 6 at the site of Tortuguero in the Mexican state of Tabasco. Monument 6 contains the only ancient Maya text thus far discovered that refers unequivocally to the 2012 date. The legible parts of this extremely brief text say that "the thirteenth pik will be finished [on] Four Ahaw, the third of K'ank'in. [An undeciphered event] will occur. [It will be] the descent of *Bolon Yokte K'u* (Nine Foot Tree God) to [undeciphered place]." This explicit referral to the end of the thirteenth pik occurring on the day 4 Ahaw negates prominent Swedish researcher Carl Johan Calleman's claim for an October 28, 2011 "end of era" date, because the text points exclusively to the December 21, 2012, date that falls on 4 Ahaw in the 260-day *cholq'ih* calendar. More significantly, the primary event predicted for that day—the arrival of the deity Bolon Yokte K'u—is suggestively similar to some currents in 2012 ideology. John Major Jenkins has pointed out that the god Bolon Yokte K'u has strong connections to both warfare and Maya creation mythology.[18] Jenkins further suggests that this deity can sometimes take the iconographic form of the World Tree, a symbol of the Milky Way and a key astronomical feature in ancient Maya creation lore. The Monument 6 hieroglyphs unfortunately provide us with extremely little information concerning 2012, and not one of the thousands of other ancient Maya texts even mentions the date. Even so, we at least know with certainty from this tiny passage that some ancient Maya actually did, in fact, see 2012 as significant and that this 13 pik date had associations with both warfare and creation events, occurrences that presumably could herald the dawn of a new era.

The only other historical source of authentic Maya predictions possibly related to 2012 is the collection of colonial-period texts from the Yucatán Peninsula known as the books of *Chilam Balam*. While the cryptic nature of these deeply esoteric texts has lent itself to highly creative speculative manipulation by several New Age teachers, the *Chilam Balam* books do, in fact, contain at least one minor reference that has a legitimate, if weak, link to the 2012 date. The books describe oracular associations with a series of successive k'atun periods, each measuring 7,200 days, slightly less than 20 years. The text states that "4 Ahaw k'atun is the eleventh k'atun according to the count. Chichen Itza is the seating of the k'atun. The settlement of the Itzas comes. The quetzal comes, the green bird comes. He of the yellow tree comes. Blood-vomit comes. K'uk'ulkan shall come." Yukatek Maya oracle-scribes named k'atuns in accord with the period's closing date in their 260-day ritual calendar. The current k'atun, the one ending in 2012 on the Northern Hemisphere winter solstice, is a 4 Ahaw k'atun, since it ends on the day 4 Ahaw. However, k'atuns named 4 Ahaw recur approximately every 256 years, and the *Chilam Balam* text explicitly ties this prophecy to the immediately previous four Ahaw k'atun in the middle of the eighteenth century. Still, many scholars believe this oracular text refers primarily to the arrival of a historical figure known as K'uk'ulkan in the Yucatán in a far earlier 4 Ahaw k'atun that ended in November 987 C.E. However, within the context of a Maya temporal orientation that focuses on cyclical repetition, any 4 Ahaw k'atun would share similar qualities. Thus, there is a remote possibility that the passage refers to the 4 Ahaw k'atun ending in 1500 C.E., around the time of the Spanish invasion into the Maya world and the numerous diseases they brought with them. Presumably, since the December 21, 2012, date closes a 4 Ahaw k'atun, the text implicitly could refer to the current k'atun cycle, even though its more obvious connections to the earlier k'atun periods in Maya history are significantly stronger. In the contemporary context, these colonial Maya references to violence and sickness—plus the green renewal image of the transforming feathered serpent, K'uk'ulkan—echo the cataclysmic and regenerative motifs at the heart of the 2012 phenomenon.

Apart from this tangential reference in the *Chilam Balam* books and the single, short hieroglyphic passage on Tortuguero's Monument 6, there are no other historical Maya texts that clearly refer to 2012. Even the sacred *Popol Vuh*, arguably the most sacred of all Maya texts, makes no mention of the year 2012 or of future world change. However, several contemporary Maya communities make substantive references analogous to core currents in the 2012 movement. While these sources are practically unknown in the 2012 phenomenon, these authentic indigenous beliefs are similar in that they point to an approaching period of significant, even catastrophic, world change and subsequent renewal, but without explicit reference to the year 2012. For example, within the prophetic tradition of the Cruzo'ob Maya in the Mexican state of Quintana Roo, there has long existed a belief in a rapidly approaching cataclysmic period of warfare that will lead to the destruction of their current world and the creation of a new Maya society. In 1930, for example, one Maya man in the village of Chemax declared,

> *I guess I'll tell you the story of beautiful holy Lord for you to hear,*
> *because I have read the testament [a sacred almanac] of beautiful*
> *holy Lord, where he says [that in] 2000 and a few more years it*
> *will end on earth. But if they have been very good Christians on*
> *earth, he will not end it. . . . He begins to diminish, beautiful holy*
> *Lord, His merciful grace [corn], the end of the road.*[19]

In a similar vein, according to Paul Sullivan, in 1989, Maya of the Cruzo'ob village of Xcacal Guardia believed that war was "not much further off than the year 2000 and that it is inevitable."[20] One Mexican anthropologist familiar with the Cruzo'ob communities states that "the wait for fulfillment of the promises of the Cross, the imminent arrival of the end of the world, and the creation of a new Mayan society are dynamic elements that are present, strong, alive, and active."[21] The son of an elder Cruzo'ob Maya prayermaker stated,

> *The end of the world is going to arrive in the year 2000 and a*
> *bit more, but no one knows how much a bit is. It might be thirty*
> *years or a century, nobody knows when it will be. . . . The legend*
> *used to say that in Chan Santa Cruz there would arrive a time in*
> *which four white ropes would cross; these ropes are the highways*
> *to Merida, Vallodolid, Cancun, and Vigia Chico. These ropes are*
> *a sign that time is going to change; or perhaps that time already*
> *changed. When times change, a new war is also near.*[22]

I recently visited with prayer makers and guardians at the five principal Cruzo'ob shrine sites in the Yucatán, including the new one in Felipe Carrillo Puerto.[23] While not a single person there was aware of any predicted events connected explicitly with the 2012 date, nearly every individual I spoke with made quick reference to an imminent period of severe difficulties that might include warfare. Without exception, they emphatically avoided predicting a precise date for the cataclysm, insisting that "only God knows" when it will occur. The Cruzo'ob Maya dream of an apocalyptic end to Hispanic domination of their world seems a natural consequence of their historical circumstances. They have suffered severe injustices, first at the hands of the Spaniards and then the Mexicans, including a full-fledged campaign of extermination during the nineteenth-century Caste War. Coincidentally, Maya hopes for world renewal mirror those at the heart of the New Age 2012 movement.

Similarly, among the few remaining Lacandon Maya traditionalists in Chiapas, there is a belief that the end of the world is at hand. In 1978, the late, well-known spiritual leader of Najá, Chan K'in, stated, "Our Lord Hachakyum will make everything die. . . . The grass wishes to die. The seed, the animals all wish to die. And the True People also [the Lacandon]—we all die. In thirty years Hachakyum will destroy the world."[24] The tradition of world end and renewal is so strong among the Hach Winik (Lacandon) that they have a special word for it, *xu'tan*. Even though nearly all of the Hach Winik in all three of their principal settlements have converted to various Christian denominations, recognition of the term "xu'tan" persists. Several Lacandon men recently made clear to me that the Christian Apocalypse they have learned about in their local churches and the Maya xu'tan are not the same thing. In particular, xu'tan implies the cutting, burning, and destruction of vegetation. As with the Cruzo'ob Maya beliefs concerning world renewal, those of the Lacandon reflect their own peculiar historical circumstances as witnesses to the virtual destruction of their traditional homeland, the *selva lacandona*, once one of the largest rain forests on the American continent. The living green world that nurtured Lacandon culture for hundreds of years is now almost completely gone. Having witnessed this ongoing deforestation, the Lacandon belief in impending world destruction seems quite logical. Coincidentally, the Lacandon beliefs also parallel the cataclysmic dimensions in the 2012 movement.

Recently, a Mam Maya traditionalist in Todos Santos Cuchumatán, Guatemala, drew my attention to prophetic traditions that persist among highland Maya that also augur a period of epochal change. He said that he believes the Maya are approaching the culminating period of a divine trial in which they have had to endure nearly 500 years of oppression under the *mos*,[25] Guatemala's nonindigenous population. He stated that the Maya are currently witnessing a process of purification that will prepare them for what he called, in Mam, the *Ak'ah Sqixa,* or New Dawn. According to his understanding, the light of this New Dawn will return the Maya to their rightful place as rulers of their own lands.[26] Once again, as with the Cruzo'ob and Lacandon prophetic traditions, the world renewal traditions among the Mam, K'iche', Kaqchikel, and other Maya in highland Guatemala tend to reflect their own particular historical circumstances; in this case, their brutally harsh experience in the region since the arrival of the first Spaniards in the early 1500s.

Balance is a core Maya value. Just as Maya culture seeks balance between hot and cold, between male and female, its orientation toward equilibrium implies that there will inevitably be a reversal of the centuries of repression suffered by highland Maya under the rule of the Spanish-speaking minority.

Because use of the Long Count on which the 2012 date is based ended many centuries ago, specific references by contemporary Maya to the date are rare and appear only among those recently exposed to the Long Count through formal or informal academic study. Gaspar González, a Maya novelist and prominent cultural activist, is one of those few Maya who have spoken explicitly about the year 2012. González contextualized the recent Guatemalan civil war as part of an extended period of severe tribulation that would prepare his people for the next cycle in the human experience. He has said that the current age of the human beings made of corn was ending, and that after 2012, there will be a societal rebirth into what he called "*una nueva era de la luz*" (a new age of light). Recently, González added the following comments:

> *From the perspective of contemporary Maya, 2012 constitutes a very important point in the history of humanity since time is a variable that greatly influences the life of the planet and everything that exists on it. Human beings do not exist by coincidence or by a work of chance. They are part of a plan to carry out a mission in this part of the universe. The world is still not totally finished in its creation and perfection; this human creature has a role to play in the world and its preservation. One could say that the life of the planet depends on human beings and what they do in their existence.*[27]

As a senior member of the cultural revival movement known as the *movimiento maya*, a serious scholar, a native speaker of Q'anjob'al Maya, and a member of the Guatemalan Academy of Maya Languages, González has impeccable credentials as a Maya spokesperson. His specific references to the year 2012 are undoubtedly a result of academic study of the once-forgotten

Maya Long Count calendar. They coincide perfectly with pre-existent prophetic currents—such as those among the Hach Winik, Cruzo'ob, and highland Maya—that are not, however, explicit in terms of date.

Another Maya intellectual, the Jakalteko novelist, professor, and cultural activist Victor Montejo, echoes González's emphasis on the active role of human beings in the coming age, after 2012:

> *Prophetic expressions of the indigenous peoples insist on the protagonist role that new generations must play at the close of this Oxlanh B'aktun [thirteen B'aktun] and the beginning of the new Maya millennium. The ancestors have always said that "one day our children will speak to the world.". . . This millennial or b'aktunian movement responds to the close of a great prophetic cycle . . . the great prophetic cycle of four hundred years in the Maya calendar. For the Maya, this is not the close of the second millennium or 2,000 years after Christ, but rather the close of the fifth millennium according to the ancient Maya calendar initiated in the mythical year that corresponds to 3114 B.C. . . . The b'aktun includes the global concept of time and the regeneration of life with new ideas and actions. In other words, the theoretical b'aktunian approach leads us to understand the effect of human ideas and actions on all that exists on the earth and their effects on the environment and cosmos.*[28]

In recent years, knowledge of the year 2012 has spread beyond the realm of internationally known Maya spiritual teachers and Maya intellectuals to include a few spiritual guides still living and working within their highland Guatemalan communities. Because the year's significance arises from within the context of the ancient dimensions of their culture, their tendency is to incorporate it into long-existing prophetic traditions and hopes for a revival of Maya culture and indigenous political power. The K'iche' spiritual guide Rigoberto Itzep quotes community elders in Momostenango, who say, "You will still see many warnings. You will still see and hear strange things. You will still see great ruin. There will be many changes on Earth." According to Itzep, these K'iche' Maya elders never specified the year 2012; but for him, the potential for the year is obvious: "The ideological power of the West in its entirety might expire forever in 2012"—words that undoubtedly represent the heartfelt wish of many Maya in the region.

But even for Itzep, and other traditional Maya like don Alejandro, the exact date in 2012 is not especially critical. Instead, they view the date simply as a temporal marker in the midst of vast cyclical processes that were set in motion long ago. As José María Tol Chan, an *ajq'ij* from highland Guatemala, recently told me, "It is an event that has already begun; there are already signs. . . . Humans more than ever should pay close attention to all the events that disturb balance. There are teachings that we living beings should extract from the stages through which we pass. It's not that we are arriving at a zero hour in 2012; it's already beginning."[29]

Even though prophecies of world renewal have long existed among the Maya, the idea of a coming radical transformation has entered the contemporary Maya world most broadly through the teachings of numerous fundamentalist Protestant denominations that have grown rapidly in the region during recent decades.[30] The disastrous 1976 earthquake that shook Guatemala not only killed tens of thousands of Maya, but also brought a new wave of Christian missionaries pushing their religious ideology along with much-needed relief supplies. The cataclysmic devastation of the earthquake was followed by a particularly brutal military repression of the Guatemalan Maya during the late 1970s and early 1980s, when as many as 200,000 Maya civilians perished and hundreds of thousands of others found themselves wounded or displaced.

There are several complementary explanations for the massive Maya conversion to Christian fundamentalism apart from the civil war and earthquake. When missionaries referred to the coming end of the world in accord with their own interpretation of the Bible,[31] it is easily understood how some Maya might have been inclined to believe them. With the Earth shaking beneath them, witnessing the wholesale slaughter of their families and friends, seeing the perceived degeneration of their own religious traditions, and experiencing the rapid degradation of their natural surroundings, it might be a challenge not to conclude otherwise. Christian fundamentalist groups focus on the biblical passages of the book of Revelation and, of course, do not refer to the year 2012. Although there is no evidence that Maya Christian fundamentalists will embrace the 2012 date, global communication networks make it virtually unavoidable that the two "millennial" currents will eventually come into greater contact with one another. It seems quite possible that some Maya fundamentalists who are awaiting the end of the world will see the 2012 date as an attractive way to blend their new faith with the ways of their revered ancestors.

Continued cross-pollination between contemporary Maya and New Age participants in the 2012 movement is inevitable and likely to increase, especially as more Maya spiritual teachers travel internationally and mingle with counterparts from a variety of world traditions. Many of these same Maya teachers also guide groups of New Age spiritual seekers on tours of various sites in the Maya world, often providing opportunities for participants to interact with at least some members of the local communities. There have already been events that focused on the 2012 date, in Tikal, Momostenango, and around Lake Atitlán in Guatemala, as well as in Mérida, Tulum, Coba, and elsewhere in Mexico. While most of these events are conducted primarily in English, there are now a few Spanish-language equivalents.

Apart from New Age tourism, global communication and an extensive pattern of Maya immigration to the United States will facilitate development of greater awareness concerning the 2012 date in the general contemporary Maya population. The fact that the 2012 date has authentic connections to their ancient *abuelos*—their ancestral grandparents—will no doubt attract Maya traditionalists and those in the Maya cultural revitalization movement who are attempting to resuscitate key components of their heritage. Hopes are alive among nearly all Maya for the dawn of a new era, one based on justice and respect. In accord with their cyclical view of the world, many long for a fifth creation to follow the first four, as described in their sacred *Popol Vuh*. For those with a more political agenda, the date may come to represent a culturally preordained shift point. The world's leading Maya activist, Rigoberta Menchú, for example, has already announced plans to run for the Guatemalan presidency in the year 2011 as part of a Maya political party. This may not be a coincidence. If she wins the elections, she would enter office in the year 2012 as the nation's first Maya president. Thus, even though little of the current speculation in the 2012 phenomenon has a genuine source in the Maya world, the year may turn out to be one of the most significant in Maya history. Long-standing prophetic Maya traditions linked to world renewal can now easily be associated with a specific date that has authentic roots deep in their ancient past. The region's rapid environmental degradation, evangelical

Protestantism's growing influence among Maya of apocalyptic scenarios, and an increased awareness among Maya of the 2012 phenomenon outside the region may coincide to create a self-fulfilling prophecy. These trajectories may very well lead to a period of radical change in the Maya world and, it is hoped, to a brighter future for their children and the ecosystems they call home. Even though the 2012 date may, in truth, have relatively little historical significance in Maya culture, through unrelated environmental and cultural currents, the date may take on truly transformative dimensions for contemporary Maya.

For non-Maya, the December 21, 2012, date will likely be a "nonevent" similar to what was the widely anticipated Y2K phenomenon. Fortunately, this need not keep non-Maya from reaping the benefits of Maya wisdom. While many involved with the 2012 phenomenon see the Maya as messengers from the stars, masters of time, or guardians of galactic tradition of human evolution, in truth, the Maya's most valuable message for the outside world comes from planet Earth itself. The real teachings of the Maya are the lessons they learned through a hundred generations of continuous living in intimate relationship in the same relatively small region. Such experiential familiarity with nature and the various ecosystems that have sustained the Maya for millennia have led to a collective understanding of their dependence upon what some of them refer to as "Our Mother." The traditional Maya sense of intimacy with the living world gives rise to a broadened perspective on human life that has led to an enhanced sense of awe, respect, and compassion, all fostered by intensely nurturing and engaged parenting. As a simple consequence of their culture's approach to life, many Maya still experience an ongoing shared awareness of their bonds with one another and with nature herself. These practical, hard-earned lessons in living are the true gift of Maya culture to the wider world. As the much-heralded year 2012 approaches, the real lessons of daily living in the Maya world offer us a means to enhance the quality of our own lives and the health of the planet, regardless of the significance of the date itself.

PART TWO

Science, Business, and Politics
in the Context of 2012

The Birthing of a New World

ERVIN LASZLO

*According to Ervin Laszlo, we are at a critical time in history.
Life as we know it faces a profound change: the danger of global
collapse—or the opportunity for worldwide renewal. In his book* The
Chaos Point, *Laszlo, the founder of systems philosophy and general
evolution theory, provides an overview of the decade leading up to
2012. We are nearing what he calls a "chaos point," in which we can
either choose a more sustainable world or see our civilization and
ecological systems break down. For Laszlo, one thing is certain—the
2012 chaos point will be the end of the world as we know it. In
the following excerpt from* The Chaos Point, *Laszlo describes
the challenges we face, from global epidemics and environmental
disasters to worldwide collapses in business and international
political alliances. He also outlines the benchmarks for the evolved
consciousness we are capable of attaining after 2012. Laszlo answers
this question: What changes must we make on a personal, national,
and international scale in order to avoid total chaos in 2012?*

New thinking starts with greater insight into the transformation that ushers in a new world in place of the old. But for
new thinking to be effective, we should have some idea of
what it involves. Just what kind of a process is the birthing of
a new world?

Talk of fundamental change in the world around us is often met with skepticism. Change in society, we are told, is never really fundamental: as the French saying goes, *plus ça change, plus c'est la même chose* (the more things change, the more they are the same). After all, we are dealing with humans and human nature, and these will be very much the same tomorrow as they are today. A more sophisticated variant of the prevalent view adds that certain processes in society—trends—make a significant difference as they unfold. Trends, whether local or global, micro or mega, introduce a measure of change: As they unfold, there are more of some things and less of others. This is still not fundamental change, for the world is still much the same, only some people are better off and others worse off. This view is the one typically held by futurists, forecasters, business consultants, and all manner of trend analysts. Their extrapolations are highly regarded, as the popularity of literature dealing with megatrends attests.

Governmental agencies also engage in forecasting trends. The unclassified report of the U.S. National Intelligence Council, *Global Trends 2015: A Dialogue About the Future with Nongovernmental Experts*, was published in 2000. According to this document, the state of the world in the year 2015 will be determined by the unfolding of key trends, catalyzed by key drivers. The seven key trends and drivers are demographics, natural resources and environment, science and technology, the global economy and globalization, national and international governance, future conflict, and the role of the United States. The way these trends unfold under the impact of their drivers can produce four different futures: a future of inclusive globalization, another future of pernicious globalization, a future of regional competition, or a postpolar world. The main deciders are the effects of globalization—positive or negative—and the level and management of the world's potential for interstate and interregional conflict.

When the trends unfold without major disruption, we get what the experts call "the optimistic scenario." In this perspective, the world of 2015 is much like today's world except that some population segments (alas, a shrinking minority) are better off and other segments (a growing majority) are less well-off. The global economy continues to grow, although its path is rocky and marked by financial volatility and a widening economic divide.

Economic growth may be undone, however, by events such as a sustained financial crisis or a prolonged disruption of energy supplies. Other "discontinuities" may occur as well:

- Violent political upheavals due to a serious deterioration of living standards in the Middle East (This has now happened, with dramatic consequences.)
- The formation of an international terrorist coalition with anti-Western aims and access to high-tech weaponry (now a real and growing threat)
- Rapidly changing weather patterns that inflict grave damage on human health and on economies (This is now more imminent than ever.)
- A global epidemic on the scale of HIV/AIDS
- The antiglobalization movement growing until it becomes a threat to Western governmental and corporate interests
- The emergence of a geostrategic alliance—possibly of Russia, China, and India—aimed at counterbalancing the United States and Western influence
- Collapse of the alliance between the United States and Europe
- Creation of a counterforce organization that could undermine the power of the International Monetary Fund and the World Trade Organization and thus the ability of the United States to exercise global economic leadership

In the year 2000, it was anybody's guess whether the world of 2015 would be the same kind of world or something quite different. In 2005, this is no longer an open question. The world in 2015 will be very different from what it is today—not to mention from what it was at the beginning of this century.

The National Intelligence Council, however, is still producing linear extrapolations on what the future will be like. According to another report published in early 2005, titled *Mapping the Global Future* and based on consultations with 1,000 futurists around the globe, the world in 2020 will not be very different from today. Terrorism will still be present, although the prospect of wars conducted by major powers will recede. It is a "relative certainty" that the United States will remain the most powerful nation, economically, technologically, and militarily, although a possible—but manageable—erosion of U.S. power must also be reckoned with.

Such reports highlight the limits of trend-based forecasting. They ignore the fact that trends do not only unfold in time; they can break down and give rise to new trends, new processes, and different conditions. This possibility needs to be considered, since no trend operates in an infinitely adapted environment; its present and future have limits. These may be natural limits due to finite resources and supplies, or human and social limits due to changing structures, values, and expectations. When a major trend encounters such limits, the world is changing and a new dynamic enters into play. Extrapolating existing trends does not help in defining the emerging world.

To know what happens when a trend breaks down calls for deeper insight. It calls for going beyond the observation of current trends and following their expected path—it requires knowing something about the developmental dynamics of the system in which the observed trends appear and may disappear.

Such knowledge is provided by modern systems theory, especially the branch popularly known as "chaos theory." Because of the unsustainability of many aspects of today's world, the dynamics of development that will apply to the future is not the linear dynamics of classical extrapolation, but the nonlinear chaos dynamics of complex-system evolution.

THE DYNAMICS OF TRANSFORMATION:
A BRIEF EXCURSION INTO CHAOS THEORY

At the dawn of the twenty-first century, we can no longer ignore that current trends are building toward critical thresholds, toward some of the famous (or infamous) "planetary limits" that in the 1970s and 1980s were said to be the limits to growth. Whether they are limits to growth altogether is questionable, but they are clearly limits to the *kind* of growth that is occurring today. As we move toward these limits, we are approaching a point of chaos. At this point, some trends will deflect or disappear, and new ones will appear in their stead. This is not unusual: Chaos theory shows that the evolution of complex systems always involves alternating periods of stability and instability, continuity and discontinuity, order and chaos. We are living in the opening phases of a period of social and ecological instability—at a crucial decision window. When we reach the point of chaos, the stable "point" and "periodic" attractors of our systems will be joined by "chaotic" or "strange" attractors. These will appear suddenly—as chaos theorists say, "out of the blue." They will drive our systems to the crucial point where it will select the one or the other of the paths of evolution available to it.

In the current decision window, our world is supersensitive, so that even small fluctuations produce large-scale effects. These are the legendary "butterfly effects." The story goes that if a monarch butterfly flaps its wings in California, it creates a tiny air fluctuation that amplifies and amplifies and ends by creating a storm over Mongolia.

The discovery of the butterfly effect is linked with the art of weather forecasting, having its roots in the shape assumed by the first chaotic attractor discovered by U.S. meteorologist Edward Lorenz in the 1960s. When Lorenz attempted to computer-model the supersensitive evolution of the world's weather, he found a strange evolutionary path, consisting of two different trajectories joined together like the wings of a butterfly. The slightest disturbance would shift the evolutionary trajectory of the world's weather from one wing to the other. The weather, it appears, is a system in a permanently chaotic state—a system permanently governed by chaotic attractors.

Subsequently, a considerable variety of chaotic attractors has been discovered. They are applicable in some measure to all complex systems, above all to living systems. Living systems are remarkable systems; they do not move toward equilibrium, as classical physical systems do, but maintain themselves in their improbable state far from chemical and thermal equilibrium by constantly replenishing the energies and matter they consume with fresh energies and matter flowing from their environment. (Physicists say that they balance the positive entropy they produce internally by importing negative entropy from their surroundings.)

Humans, as other complex organisms, are supersensitive dynamic systems permanently at the edge of chaos, as are the ecologies and societies formed by living systems. These collective systems are wider and more enduring than their individual members, but the dynamics of systems evolution also applies to them.

The evolution of individual and collective organic systems can usually be described with differential equations that map the behavior of the systems in reference to the principal system constraints. This is not feasible in regard to the societies formed by human beings; here the presence of mind and consciousness complicates the evolutionary dynamics. The consciousness of its human members influences the system's behavior, making it far more complex than the behavior of nonhuman systems.

In periods of relative stability, the consciousness of individuals does not play a decisive role in the behavior of society, since a stable social system dampens deviations and isolates the deviants. But when a society reaches the limits of its stability and turns chaotic, it becomes supersensitive, responsive even to small fluctuations, such as changes in the values, beliefs, worldviews, and aspirations of its members.

We now live in a period of transformation, when a new world is struggling to be born. Ours is an era of decision—a window of unprecedented freedom to decide our destiny. In this decision window, "fluctuations"—in themselves small and seemingly powerless actions and initiatives—pave the way toward the critical chaos point where the system tips in one direction or another. This process is neither predetermined nor random. It is a systemic process that can be purposively steered.

As consumers and clients, as taxpayers and voters, and as public opinion holders, we can create the kinds of fluctuations—the actions and initiatives—that will tip the coming chaos point toward peace and sustainability. If we are aware of this power in our hands, and if we have the will and the wisdom to make use of it, we become masters of our destiny.

CHAOS DYNAMICS IN SOCIETY

The transformation of society is not a chance-ridden, haphazard process; chaos and systems theory disclose that it follows a recognizable pattern. Typically the transformation manifests four major phases.

1. *The Trigger Phase.* Innovations in "hard" technologies (tools, machines, operational systems) bring about greater efficiency in the manipulation of nature for human ends.

2. *The Accumulation Phase.* Hard technology innovations change social and environmental relations and bring about successively
   - Higher levels of resource production
   - Faster growth of population
   - Increasing societal complexity
   - Increasing impact on the social and the natural environment

3. *The Decision Window.* Changed social and environmental relations put pressure on the established order, placing into question time-honored values and worldviews, and the ethics and ambitions associated with them. Society becomes unstable, supersensitive to all fluctuations.

4. *The Chaos Point.* Here the system is critically unstable. The status quo becomes unsustainable and the system's evolution tips in one direction or another:

   a. Evolution ("Devolution") Toward Breakdown. The values, worldviews, and ethics of a critical mass of people in society are resistant to change, or change too slowly, and the established institutions are too rigid to allow for timely transformation. Inequity and conflict, coupled with an impoverished environment, create unmanageable stresses. The social order degenerates into conflict and violence.

   or

b. Evolution to Breakthrough. The mind-set of a
critical mass of people evolves in time, shifting
the development of society toward a more
adaptable mode. As these changes take hold, the
improved order—governed by more adapted values,
worldviews, and ethics—establishes itself. The
economic, political, and ecological dimensions of
society stabilize in a nonconflictual and sustainable
mode.

We now look at the four-phase process of societal transformation in the context of the contemporary world.

1. The Trigger Phase, 1800–1960. Innovations in "hard"
technologies (tools, machines, operational systems) trigger fundamental changes in the way people live and work for the sake
of creating greater efficiency in organizing people, resources,
and nature to reach the desired ends.

Until the second half of the eighteenth century, the 8,000
years that separated the rise of the Neolithic from the advent of
the Industrial Age saw relatively few fundamental technological
innovations. Basic agricultural tools were refined but not substantially modified: The sickle, hoe, chisel, saw, hammer, and
knife continued in use in substantially unchanged forms. More
radical changes occurred only in regard to the technologies of
irrigation and the introduction of new varieties of plants.

By the year 1800, the Industrial Revolution had been brew-
ing in England for about half a century. Yet even in the three
countries at the forefront of the revolution—England, France,
and the United States—there were no telegraphs, railroads,
macadamized roads, or steamboats. The iron and steel industries were still embryonic. But the steamboat was invented in
1802, and oil was found in Pennsylvania in 1859. In the middle

of the nineteenth century, the Industrial Revolution entered into full swing, bringing an entire battery of new technologies onto the scene.

The first breakthroughs occurred in textiles: Innovations in spinning cotton stimulated related inventions, leading to machines capable of factory-based mass production. Industrial development soon spread from textiles to iron, as cheaper cast iron replaced more expensive wrought iron.

Closely following on the heels of innovations in the machine tool industry were developments in the chemical industry. Many of the twentieth-century technologies in the automobile, steel, cement, petrochemical, and pharmaceutical industries were spawned in the 1860s and the years that followed. Modern steel mills are, for the most part, still based on the Bessemer steel process developed at that time; the rotary kiln, patented by Fredrick Rancome in 1885, is still used in today's cement production; and the synthetic dyes of the late eighteenth century were basic to the development of modern chemical industries. The traction-based combustion engine, a key innovation in modern transportation, appeared in the 1880s, simultaneously with Edison's electric lightbulb, followed by Marconi's wireless, and the Wright brothers' flying machine.

In the first half of the twentieth century, these technological innovations shifted industrial production from coal and steam, textiles, machine tools, glass, pre-Bessemer forged steel, and labor-intensive agriculture to electricity, the internal combustion engine, organic chemistry, and large-scale manufacturing.

2. The Accumulation Phase, 1960 to the Present. Hard technology innovations accumulate and irreversibly transform social and environmental relations. They bring about, at an ever-increasing rate,

- Higher levels of resource production

- Faster growth in the population
- Increasing societal complexity
- Increasing impact on the natural environment

In the early 1960s, nearly 160 years after the innovations that led to the unfolding of the first Industrial Revolution, a new type of technological innovation occurred. The "second Industrial Revolution" replaced reliance on massive energy and raw material inputs with the more intangible resource known as *information*. In the last quarter of the twentieth century, a rapidly growing quantity of information came to be stored on optical disks and communicated via fiber optics, with computers equipped with sophisticated programs elaborating the data. The new "soft" technologies made the classical "hard" technologies more efficient. Sophisticated information systems rationalized and dropped the cost of production and consumption and led to vast increments in the mining, production, use, and, ultimately, discard of the manufactured goods produced by ever-more-powerful automated or semi-automated technologies.

The spread of industrial technologies to the four corners of the globe produced a series of profound transformations, globalizing the economic and financial sectors while leaving social structures locally diverse and disparate. For a minority, it brought new wealth and great increases in the material standard of living, but for the growing masses, it brought deepening poverty and seemingly hopeless marginalization. Uneven and imbalanced globalization sparked a new gold rush for the wealth promised by the high-technology service and production sectors. The unreflective rush for wealth broke apart traditional structures and placed in question established values and priorities. It led to the exploitation, and occasionally overexploitation, of both renewable and nonrenewable resources, and it degraded the livability of the urban, as well as the rural, environment.

3. The Window of Decision, 2005–2012. The dominant social order is stressed by radically changed conditions that place in question established values, worldviews, ethics, and aspirations. Society enters a period of ferment. Now the flexibility and creativity of the people create that subtle, but all-important, "fluctuation" that decides which of the available paths of development society will hereafter take.

By the end of the twentieth century, globalization had reached a new phase: The world system had become increasingly and visibly unsustainable. In the first decade of the twenty-first century, the progressive globalization of the economy, coupled with intensifying contact among disparately developed cultures and societies, has built toward a critical epoch in which our systems have become unsustainable and increasingly sensitive to change. This state was triggered by high levels of stress, including terrorism and war, conflict in the political sphere, vulnerability in the economic arena, volatility in the financial sphere, and worsening problems with climate and the environment.

4. The Chaos Point, 2012. The processes initiated at the dawn of the nineteenth century and accelerating since the 1960s build inevitably toward a decision window and then toward a critical threshold of no return: the chaos point. Now a simple rule holds: We cannot stand still, we cannot go back, we must keep moving. There are alternative ways we can move forward. There is a path to breakdown, as well as a path to a new world.

In remarkable—and *perhaps* not entirely fortuitous agreement with the date predicted by the Mayan civilization—the chaos point is likely to be reached on or around the year 2012. The Mayan calendar indicates that the Age of Jaguar, the thirteenth baktun, or long period of 144,000 days, will come to an end with the fifth and final Sun on December 22, 2012. That date, according to the Mayan system, will mark the "gateway" to a new epoch of planetary development, with a radically different kind of consciousness.

The year 2012 is indeed likely to be a gateway to a different world, but whether to a better world or to a disastrous one is yet to be decided. At that point, alternative paths open to us:

a. *The Breakdown Path: Devolution to Disaster.*
Rigidity and lack of foresight lead to stresses that the established institutions can no longer contain. Conflict and violence assume global proportions, and anarchy follows in their wake.

or

b. *The Breakthrough Path: Evolution to a New Civilization.* A new way of thinking with more adapted values and more evolved consciousness mobilizes people's will and catalyzes a fresh surge of creativity. People and institutions master the stresses that arose in the wake of the preceding generation's unreflective fascination with technology and untrammeled pursuit of wealth and power. By the year 2025, a new era dawns for humanity.

The insight we get from this four-phase transformation dynamic is simple and straightforward. In society, fundamental change is triggered by technological innovations that destabilize the established structures and institutions. More adapted structures and institutions await the surfacing of a more adapted mindset in the bulk of the population. Thus in the transformation of our world, technological innovation is the trigger. The decider, however, is not more technology, but the rise of new thinking—new values, perceptions, and priorities—in a critical mass of the people who make up the bulk of society.

You possess a more evolved consciousness when you:

1. Live in ways that enable all other people to live as well, satisfying your needs without detracting from the chances of other people to satisfy theirs.

2. Live in ways that respect the right to life and to economic and cultural development of all people, wherever they live and whatever their ethnic origin, sex, citizenship, station in life, and belief system.

3. Live in ways that safeguard the intrinsic right to life and to an environment supportive of life for all the things that live and grow on Earth.

4. Pursue happiness, freedom, and personal fulfillment in harmony with the integrity of nature and with consideration for the similar pursuits of others in society.

5. Require that your government relate to other nations and peoples peacefully and in a spirit of cooperation, recognizing the legitimate aspirations for a better life and a healthy environment of all the people in the human family.

6. Require business enterprises to accept responsibility for all their stakeholders as well as for the sustainability of their environment, demanding that they produce goods and offer services that satisfy legitimate demand without impairing nature and reducing the opportunities of smaller and less-privileged entrants to compete in the marketplace.

7. Require public media to provide a constant stream of reliable information on basic trends and crucial processes to enable you and other citizens and consumers to reach informed decisions on issues that affect your and their lives and well-being.

8. Make room in your life to help those less privileged than you to live a life of dignity, free from the struggles and humiliations of abject poverty.

9. Encourage young people and open-minded people of all ages to evolve the spirit that could empower them to make ethical decisions of their own on issues that decide their future and the future of their children.

10. Work with like-minded people to preserve or restore the essential balances of the environment, with attention to your neighborhood, your country or region, and the whole of the biosphere.

Addressing a joint session of the U.S. Congress in February of 1991, Václav Havel, then the president of Czechoslovakia, said, "Without a global revolution in the sphere of human consciousness, nothing will change for the better . . . and the catastrophe toward which this world is headed—the ecological, social, demographic, or general breakdown of civilization—will be unavoidable." Havel's point is well taken, but it is not a reason for pessimism: The breakdown of civilization can be avoided. Human consciousness can evolve. It is already evolving, and you can help it evolve further.

Former U.S. president Harry Truman remarked in turn, "The buck stops here," meaning the desk at the Oval Office. Today the buck is more democratic: It stops not only in the White House, but with you and me and everyone around us. It comes in the form of a challenge: Reexamine your thinking, evolve your consciousness. If you do so, the brave but as-yet insufficiently powerful and organized movement toward a more holistic, peaceful, and sustainable civilization could turn into a groundswell that does away with the mind-set that generated our problems, and orients you and those around you toward a world you can live in and can leave with good conscience to your children.

Getting to 2012

BIG CHANGES AHEAD

JOHN L. PETERSEN

John L. Petersen, president of the Arlington Institute, is considered by many to be one of the most informed futurists in the world. He is best known for his thinking and writing about high-impact surprises—wild cards—and the process of surprise anticipation. In the following essay, he tackles the fast-approaching era of 2012 by assessing natural resources, species extinction, climate change, economic ramifications, and the necessity for a new paradigm. He answers such questions as, How can we avoid the seemingly inevitable issues of finite resources and the misuse of those resources? What does this new paradigm in 2012 look like?

Consider this recent BBC headline: "Current global consumption levels could result in a large-scale ecosystem collapse by the middle of the century, environmental group WWF has warned." One that followed was, "Climate change threatens supplies of water for millions of people in poorer countries, warns a new report from the Christian development agency Tearfund."

About the same time, the *Washington Post* said, "Birds, bees, bats, and other species that pollinate North American plant life are losing population, according to a study released yesterday by the National Research Council."

Reuters added, "Failing to fight global warming now will cost trillions of dollars by the end of the century even without counting biodiversity loss or unpredictable events like the Gulf Stream shutting down."

Author James Howard Kunstler chimed in: "The Long Emergency is going to be a tremendous trauma for the human race. We will not believe that this is happening to us, that 200 years of modernity can be brought to its knees by a worldwide power shortage. The survivors will have to cultivate a religion of hope—that is, a deep and comprehensive belief that humanity is worth carrying on."

Then, in a landmark report compiled by Sir Nicholas Stern for the British government, comes the admonition, the world has to act now on climate change or face devastating economic consequences. Sir Nicholas estimated that, at most, humanity has 10 years before the shift is unrecoverable.

What's going on here? What does this all mean? These are extraordinary statements about massive Earth changes. Are they just random trends that happen to be showing up coincidentally? Or perhaps they reflect some big, historic underlying dynamic—maybe the world is about to experience a shift unlike anything ever seen before.

There are reasons to believe the latter could be the case. Many sources, both conventional and unconventional, suggest that we are living in a special time—that between now and 2012 the world will undergo an epochal shift to a new era. This rapid evolution will produce a world that operates in fundamentally different ways than it has in the past.

The indicators are there. Take a closer look at what is already happening.

DEMOGRAPHICS

Nearly half of all the people on the planet are under the age of 25.[1] That's the largest youth generation in history. The overwhelming majority of these young people live in the developing world, and almost a quarter are surviving on less than one dollar a day.[2] Most of them know about the quality of life in the West. Many have seen and used a computer or a mobile phone.

PEAKING OF THE GLOBAL OIL SUPPLY

Despite the increased awareness that our oil resources are finite, demand for oil is growing, significantly driven by the needs of China and India. In the past few years, it grew from 79.8 (2003) to 84.3 (2005) million barrels per day.[3] Even if the Chinese economy were to slow down, the growth is still likely to continue with a pressure from India.

Supply, on the other hand, appears to have peaked. World oil production has retreated from its all-time high of just over 85 million barrels a day achieved in December 2005 (just as geologist Kenneth Deffeyes of Princeton University had predicted). In 2006, production remained in the 84-million-barrels-a-day range every month, while demand exceeded that.[4]

"Oil curse" is a term coined to reflect the desperate situation of many oil-rich but otherwise underdeveloped countries. The Chinese are now involved in a comprehensive international outreach to African countries, buying up resources (not just oil) in Nigeria, Angola, Congo, and Sudan. So far, oil importers have used mostly economic and political means to compete for oil, but they will inevitably resort to military strategies as soon as they realize that they have passed peak oil threshold.

The report *Peaking of World Oil Production: Impacts, Mitigation, and Risk Management,* prepared by Science Applications International Corporation (SAIC) for the U.S. Department of

Energy, concludes that humanity is facing asymmetric risks associated with the peaking of oil production. Although mitigation actions initiated prematurely may result in a poor use of resources, late initiation of mitigation may result in severe consequences. Early mitigation measures are necessary to install production capacities for alternative energy in time for the peaking of oil production.

SPECIES EXTINCTION

Despite an avowed reverence for life, human beings continue to destroy other species at an alarming rate, rivaling the great extinctions of the geologic past. In the process, we are foreclosing the possibility of discovering the secrets they contain for the development of new life-saving medicines and of invaluable models for medical research, and we are beginning to disrupt the vital functioning of ecosystems on which all life depends. We may also be losing some species so uniquely sensitive to environmental degradation that they may serve as our "canaries in the coal mines," warning us of future threats to human health.[5]

An example of this is a new study that shows that the oceans' fish are being depleted so fast that eating seafood might become just a memory in 40 years. The researchers say more is at stake than our diet, for they find the dwindling of fish stocks has hurt the world economically and the ocean environmentally. Researchers say it is not too late to reverse the trend.[6]

According to Millennium Ecosystem Assessment, the challenge of reversing the degradation of ecosystems while meeting increasing demands for economy-driven services could be met partially under some scenarios that they have considered, but these involve significant changes in policies, institutions, and practices that are not currently under way.

The speed of species extinction has forced scientists to refer to the current era as the sixth extinction event, comparable to only five other events in the known history of the biosphere, the best-known example being the rapid elimination of the dinosaurs. (That's a few billion years.)

CLIMATE CHANGE

Earth is already as warm as at any time in the last 10,000 years and is within 1 degree Celsius of being the hottest it has been in a million years. If another decade goes by with business-as-usual carbon emissions, it will probably be too late to prevent northern ecosystems from triggering runaway climate change.[7]

Feedback loops (the self-reinforcing relationships among carbon dioxide emissions, global warming, and other factors) are driving the dynamics of climate change. In fact, they are the source of exponential rates of growth in surface heating. We may be entering a phase in which global warming becomes a runaway train.

Glaciers in the Himalayas are receding faster than in any other part of the world, and if the present rate of retreat continues, they may be gone by 2035. More than 2 billion people—a third of the world's population—rely on the Himalayas for potable water.[8]

An increase in global temperatures can interfere with the workings of the ocean conveyor belt and bring another ice age to Europe. The Earth's ocean system is characterized by thermal inertia. This means that it adapts slowly to global cooling and warming, but once it starts to warm up or cool down, the process extends for a long period of time. For us, it means that even if all human emissions were to stop now, the ocean's thermal inertia could sustain an increase in global temperatures.

According to conclusions made by the Intergovernmental Panel on Climate Change of the World Meteorological Organization (WMO) and the United Nations Environment Programme (UNEP), there is new and stronger evidence that most of the warming observed over the last 50 years is attributable to human activities.

MAJOR ECONOMIC DISRUPTION

During 2003–2004, concerned about possible "deflation," the U.S. Federal Reserve ran interest rates so low (1 percent) that mortgage lenders began offering below-prime mortgages with little or no money down. Refinancing of existing mortgages was at an all-time high, resulting in huge increases in mortgages (more than five times the amount between 2002 and 2006 than in the preceding five-year period). Many, if not most, of those loans (whose real interest rates were higher than "prime" mortgages secured in historical ways) had extra-low payments in the loan's early years, with a substantial increase in payments when the loan "balloons." People whose income would never have allowed them to own a home previous to that time were now buying homes; and those least able to pay off their loans began using credit cards to make up for the shortfall in income.

In 2005, for the first time since 1933, the savings rate in America became negative. This happened about the same time that personal credit card debt reached its highest level ever (the number of credit cards used in the United States grew 75 percent from 1990 to 2003, while the dollar amount charged increased by 350 percent). Consumer credit as a percentage of personal income has never been so high (30 percent increase since 2000 alone), and household debt as a percentage of household assets is also at a record high.

Independent analysis shows that credit card defaults begin about 24 months after a borrower has fundamentally overextended him- or herself. Therefore, history suggests that we should see a dramatic increase in mortgage defaults starting in 2006. In fact, the percentage of U.S. subprime loans that were made in 2006 and delinquent in payments by 60 or more days by August of 2006 rose 100 percent over similar loans made in 2005.[9] All banks have a large percentage of their assets tied up in mortgage-based securities. If the default rate on mortgages increases significantly, it could well translate into a major threat to the solvency of many banks.

The debt situation will be exacerbated by the retirement of the baby boomers and by the implementation of the new banking regulations under the auspices of the International Convergence of Capital Treasurement and Capital Standards (Basel II), which demand the revamping of the global banking system, for which no banks are likely ready.

According to Warren Buffet, the current financial system is highly unstable. Highly complex financial instruments—derivatives—are time bombs and "financial weapons of mass destruction. . . . Derivatives generate reported earnings that are often wildly overstated and based on estimates whose inaccuracy may not be exposed for many years. . . . Large amounts of risk have become concentrated in the hands of relatively few derivatives dealers . . . which can trigger serious systemic problems." Derivatives can push companies into a "spiral that can lead to a corporate meltdown."[10]

Investor George Soros pronounced the same criticisms regarding the global financial system. He believes that unless fundamental reforms are implemented, the current system will continue on a spiral of crises.[11]

WATERSHED TIME?

Is the nexus of these forces a unique watershed time that will usher in a new era on this planet? Will the structures and institutions that we are all familiar with and depend upon struggle and even fail in the near future under the stress produced by breakdowns in multiple sectors?

Add to the preceding litany increasingly sophisticated terrorism, serious global shortages of drinking water, growing population pressures, and the possibility of other shocks like a global pandemic, and you've got a potential lineup for a major directional shift. The convergence of climate, oil, and financial trends alone could produce a "perfect storm" that reorders the future of humanity on this planet.

A NEW PARADIGM

If failure of the present system is what we're looking at, it would certainly be followed by a new paradigm. If the old system came down, a new one would evolve that attempted to bypass the systemic frailties of the previous world. It would necessarily be a fundamentally different way of understanding reality, attended by new perspectives of science, ecology, economy, cosmology, governing, agriculture, and education, among the other basic intellectual structures that support human activity.

The new world, as with all paradigm shifts, would not make much sense from our present perspective. Never having seen a group larger than a clan, a hunter-gatherer contemplating the future would have been hard pressed to envision a world that included people living in towns and villages. Similarly, the future that may arrive with 2012 would necessarily seem strange in the context of our upbringing.

As we get closer to the time of this epic shift, the early outlines of a new future appear to be emerging. First of all, the world is a highly interdependent and connected one. The complexity of our present communication systems links individual humans in ways that would have seemed impossible just two decades ago. (The World Wide Web had not yet been invented only 20 years ago.) As the ability to interact in increasingly more sophisticated ways develops, a point will be passed when humanity begins to act like an organism, rather than unrelated individuals and small groups. Ideas will transit the world like rumors in a small town. Concepts and perspectives will infect the global brain and produce behavior never before seen. We will see our future tied to others—many thousands of miles away from us—in ways that would have made no sense just five years ago. We will rapidly become planetary citizens.

Similarly, ecological interconnectedness is also rapidly becoming obvious. Many of us now know that we are all related to the larger environmental system in which we live in ways that we never previously understood. The death of one-third of the coral reefs in the world and 10,000 other species a year will surely affect the system that supports human life—and certainly not beneficially.

All of this new knowledge will, of necessity, change our behavior in the future. We will see ourselves as an integral part of the system in which we live. We will know that we are all in the same lifeboat, and each of our futures is a function of the future of all of us. Self-interest and security, whether characterized in personal or national terms, will very quickly encompass far more space and people than it has in the past. In the face of rapid climate change, for example, national security will approach becoming synonymous with global security.

We'll also see ourselves connected in spiritual terms. Perhaps this is where the real paradigm shift will take place. More and more individuals are beginning to experience and internalize the fact that we are connected to each other and to animals, plants, and even the Earth in ways that, though inexplicable, are nevertheless demonstrable. Serious science-based books are now being written about how human behavior is connected to the larger cosmos and how, throughout history, the cosmos has been predictably reflected in how we behave. Agricultural systems are in place that claim to tap into elemental spiritual forces that can help grow crops better. Many studies now show that the intentionality of prayer significantly affects single-cell life as well as humans; it doesn't matter whether either party knows the other and that the other is praying. There is evidence now that somehow humans anticipate big disruptions to the system (like 9/11 and the Indonesian tsunami of 2005) and begin to have extraordinary precognitive dreams before these major events.

This spiritual awareness seems to be on a trajectory that will expand to include the ability to tap into the global collective unconsciousness and may even become somewhat predictive—marrying advanced knowledge technology with dreaming and other intuitive processes.

Growing numbers of thoughtful people are coming to the conclusion that intentionality directly shapes reality. How our thoughts translate into the reconfiguration of matter and different behavior in others is not clear, but for many, lifelong experience tells them that is how it works. In all of this, there appears to be an alternative dimension(s) for communication that facilitates this interconnectivity. Who knows, perhaps human telepathy may be emergent as we see ourselves more tightly committed to each other in the future. In any case, there are a great number of indicators, both historical and contemporary, that suggest that we are approaching a time of extraordinary change.

Although no one now alive has ever lived through a similar shift, the history of the planet, as we know it, suggests that these kinds of major upheavals have happened many times in the past—in fact, they are the fundamental evolutionary mechanism for the planet. Biological life moved abruptly from single-cell life to multiple-cell life after a very long period of equilibrium. Then multiple-cell life was punctuated by a radical transformation that yielded vertebrates, followed by rapid shifts to mammals, early humanoids, and then Homo sapiens. Social evolutionary punctuations continued moving hunter-gathers into villages and towns. Eventually, the printing press was invented, which enabled the Industrial Age. Perhaps the Internet represents the new communications infrastructure upon which the radically new paradigm will be built.

Perhaps we are about to experience another punctuation in the equilibrium of human evolution. Patterns from the past suggest that the time is right for another one. The question is, Are *we* ready? If the change that is forming on the horizon is anything like it appears it might be, then all humans will need to move into a new mode of living and thinking in order to survive the transition.

There will need to be a constant orientation of openness—having a wide aperture for sensing subtle indicators that point toward coming change, and being receptive to newly emerging approaches to dealing with the rapidly changing environment. If people are not open to the suggestions and ideas of others, they will necessarily falter, as no one individual will have the capability to deal with this change on his or her own. New ideas and explanations about how reality works will begin to emerge in many places; they must be openly considered and honestly evaluated. There must also be an openness to adapt—to change rapidly when it is required.

The survivors of this epochal shift will necessarily live closer to the Earth. They will know that their food does not come from the supermarket; in fact, they may well know the farmer who grows it. They will be sensitive to the Earth in ways that they perhaps previously reserved only for humans. The current movement toward "relocalization"—shifting one's life and relationships closer to a sustaining support system—will probably be rather mature.

Effectively transitioning to this new world will require envisioning it into reality. We will all need to develop a basic but coherent idea of what the new world might look like—the principles, values, structures, behaviors, and so on—and begin to carry that common picture in our minds. We need to "get together" at regular times with as many others as possible to project the new images into the space from which everything comes. We should do it as though our lives depend on it, as they probably do.

We are all blessed to live at this time of extraordinary transformation. Each in his or her own way has a special role to play in contributing to the ultimate shape and function of this new world. That's probably why we are here at this time. We should not hesitate to play our part vigorously. Time is short.

The Mystery of A.D. 2012

KARL MARET, M.D.

As we consider the repercussions of many of the 2012 predictions, it is interesting to look at them through the lens of esoteric astrology and the cycles of time. It is widely believed that we are about to enter the Age of Aquarius. In the following essay, Dr. Karl Maret, coauthor of Awakening the Dialogue of the Heart, *explains how this new age will unfold by exploring the current era in relation to political and scientific landmark dates in history. He takes a particularly close look at the cycles of time as they pertain to the Age of Aquarius and to Saturn cycles, and what that means for the United States of America, technology, health care, and science. How did similar astrological alignments affect us in the past, and what do those outcomes tell us about the coming galactic alignment?*

As humanity approaches the year 2012, more and more people are beginning to ask, What is the meaning of the Mayan calendar ending? What kinds of changes are in store for the Earth? Will we enter a new age? What shifts in our collective consciousness might we expect? I first became interested in this unusual year when I learned that the Mayan calendar, specifically the cycle of the Long Count, was coming to an end in A.D. 2012. Having been a student of cycles for many years, I soon began to see other synchronicities with this unusual year. All life is based on rhythms and cycles. Humanity has always honored these through its sacred and civil calendars, which form a vital part of every civilization. The ending of sacred calendar cycles appears to hold significance for the whole planet, especially during this time of collective awakening to our mutual interdependence. This essay explores several cosmic events and cycles also related to 2012.

It is anticipated that during the remaining years of the Piscean Age (the current Age of the Fishes), humanity will see some of the greatest breakthroughs in scientific discoveries, technological advancement, and health-care achievement, all of which may lead to a cultural renaissance and the possible emergence of a "Universal Humanity." Many limitations in contemporary human consciousness and scientific understanding have prevented this global emergence before now. These collective obstacles are beginning to disappear as a New Age of Light, a period of spiritual flowering and tolerance, begins to illumine our Blue Planet. The final years of the Piscean Age will see the forces of love become the guiding principle in human affairs. During the remaining years of the Piscean Age, all planetary citizens are invited to rediscover the importance of engaging our personal, national, and planetary "heart energies" in the service of a collective awakening to global peace and human sustainability.

Although 2012 is an important year for all of humanity, it has special relevance to America's history. The United States, being home to people from many nations, is a virtual melting pot of humanity. By 2012, 236 years will have passed since its 1776 Declaration of Independence, a time period that encompasses eight passages of Saturn around the Sun. In an astrological and mythological context, Saturn symbolizes a nation's foundation and inner purpose and plays a formative role in the chronology of its soul purpose. From this perspective, one could say that the United States, as a nation, will soon have matured in building an eight-spoke medicine wheel of its sacred destiny.

Another important period that will end in 2012 is the completion of seven 33-year cycles spanning 231 years since the planet Uranus was discovered in 1781. In October of that fateful year, the United States also won the War of Independence at the final Battle of Yorktown. Historically, 1781 is a key marker in time related to the 84-year rhythm of Uranus, the planet associated with Aquarius, the constellation of the Water Bearer and harbinger of a new age. This essay will investigate when this promised Age of Aquarius is expected to begin and what changes we might expect to see on the planet as we approach this new era in human evolution.

THE NATURE OF TIME AND CYCLES

Around a century ago, Albert Einstein proposed the General Theory of Relativity, uniting time and space into a four-dimensional reality. Einstein's theory was essentially a theory of matter, in which the physical stuff of our world was no longer seen as composed of discrete atomic bits. Instead, matter now was recognized as existing in a continuous field, one that was mapped in curved space-time. Time became the fourth dimension and was inextricably united with space. This fundamental theory of modern physics has changed our basic understanding of the universe and the nature of reality.

For Isaac Newton, time and space were absolute concepts. For him, time existed on a single axis, with events happening earlier or later along an arrow of time. We still use this operational aspect of time in everyday life. However, human beings operate in "perceptual" time, which is intimately related to human consciousness and physiology. We experience this as a feeling of "duration." A second of time is similar to the duration of a heartbeat. Our perception of light and dark cycles during a day entrains many circadian bodily rhythms. A year is based on a rhythm of four seasons and reflects the underlying motion of the Earth around the Sun.

In contrast, scientists study the equations of motion and, in them, introduce a parameter of time symbolized by the letter t. This parameter of time is then correlated with the observed motion of material objects. It is important to comprehend that in a field theory such as Einstein's General Theory of Relativity, time, as well as space, has become an unobservable parameter that has simply become part of the general field of space-time. Yet embedded in space-time is our experience of time duration and material manifestation. That is why we create calendars to mark the "passage of time" as well as to describe the movement of the heavens around us. Social time, marked through rituals, ceremonies, rhythms of work and play, weave together the fabric of civilization. Over the millennia, the Mayan civilization was just one out of many present on the planet to influence the collective human consciousness.

What was special about the Mayan culture, and how did their mathematics deal with time? The Maya marked the passage of time through a 260-day ritual year, composed of 13 months lasting 20 days each. They also had a 360-day civil calendar. These two calendars coincided every 52 civil years, encompassing 72 ritual years. In addition, the Maya marked the unfolding of time through the Long Count, lasting some 1,872,000 days, or just over 5,125 solar years. It is this Long Count that is said to be ending around December 21, 2012, having originally begun on August 12, 3113 B.C., although scholars might argue about that exact date.

Much has been written about possible implications of the ending of this time cycle, also known as the fourth world of the

Maya and Hopi Indians but as the fifth world of the Incas. The longest count of days by the Maya was a time period of 144,000 days called a baktun. The Long Count will see the completion of a cycle of 13 baktun in December 2012, with each baktun spanning a period of a little over 394 years. Interestingly, 1,440 is the number of minutes in a day.

However, the subject of this essay is not the Mayan calendar, but the connection of the year 2012 with other historical and cosmic cycles. I would like to suggest that 2012 forms a bridging year in modern history leading to the long-anticipated Age of Aquarius, the New Age of Light and universal brotherhood/sisterhood. The fairly young American nation, in the context of the older Asian and European civilizations, has the opportunity to play a seminal role in this global cultural emergence. One important step is for a wider cross section of Americans to recognize the intimate connection of American history's cycles with 2012. Indeed, the Mayan civilization flourished on the American continent and left its mysteries impressed upon the land. As a result, the ending of the Mayan calendar's Long Count is beginning to find a strong resonance in the hearts and minds of many "cultural creatives" in America.[1]

THE CYCLES OF SATURN IN U.S. HISTORY

In Western mythological thought, a basic image is that of Father Time, known to the ancient Greeks as Cronus. Our modern words "chronology" and "chronic" stem from this Greek root. Cronus was identified with the planet Saturn (as the Romans called this god), which was the outermost, slowest "wandering star" (planet) visible to the Greeks and Romans. Saturn takes just over 29.5 years to complete one orbit around the Sun. Before 2012 begins, the United States will experience for the eighth time Saturn returning to the same zodiacal position it occupied in 1776, when the Declaration of Independence was signed. During these eight Saturn returns, the American people learned in ever-greater measure to become responsible and responsive members of the community of nations. This took tremendous spiritual discipline and the cultivation of love and goodwill for other, less-fortunate peoples in the world.

A nation that has learned this path of service may be called a "world disciple," since a disciple is anyone whose actions express his or her love and goodwill for the betterment of others, something called "world goodwill." Naturally, this path of spiritual discipline is not easy and may involve times of trials, periods of uncertainty, isolationism, and uneven leadership within the world community. Even though recent events and military actions have led to a great deal of polarization in the world, it must not be forgotten that, on balance, the United States mostly acted out of a sense of global responsibility and even "manifest destiny" to assist the community of nations in building a better, more peaceful world.

Each of the eight Saturn cycles may be seen as one of the eight segments of an eight-spoke wheel, which resembles a medicine wheel of the Native Americans. In the understanding of Buddhism, such a wheel is called a "wheel of dharma," symbolizing spiritual destiny or sacred duty. In other words, 236 years—spanning eight Saturn cycles—will have passed since this country began its journey into nationhood. In 2012, the United States will be invited to stand at a point of conscious recognition to assist humanity in creating a path toward a new era—the long-anticipated Age of Aquarius.

Figure 16 gives a brief overview of the key years in U.S. history when Saturn actually returned to the celestial position it occupied at the signing of the Declaration of Independence in 1776. These eight historical periods might be seen as reflecting

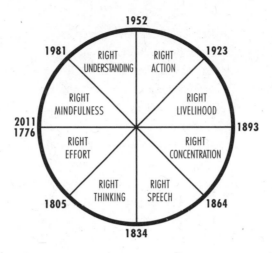

FIGURE 16: Saturn cycles, 1776–2011.

the "eightfold noble path" articulated by the Lord Buddha. These are the principles of spiritual conduct and mindful attentiveness to the collective welfare of humanity, no matter what religious tradition one follows. Seen as a complete sequence of awakened consciousness, these eight periods make up a collective wheel of spiritual awareness in the service of humanity. In 2012, the American people will have the opportunity to collectively stand as a "world disciple" and, it is hoped, again be recognized as a beacon of light and a servant leading humanity into a new era. The current challenges confronting Americans must not obscure the magnificent potential to achieve "life, liberty, and the pursuit of happiness" envisioned by our founding fathers for all people in the United States (and the world) eight Saturn cycles ago.

However, if the United States had not won the War of Independence, there would be no United States as we now know it. It is therefore of great importance to recognize the historical significance of October 19, 1781, when George Washington accepted the surrender of the British General Cornwallis at the Battle of Yorktown. At this key moment in time, the United States had truly earned her freedom and the right to become a nation. It would mark 1781 as a historical turning point for the modern world.

Earlier that year, on the evening of March 13, 1781, the German-born British astronomer Friedrich Wilhelm Herschel discovered a new celestial object, which he first thought to be a comet. It took until November of that year to confirm that Herschel had actually discovered the first new planet since antiquity. This new planet, visible only through a telescope, would later be called Uranus. It is interesting to note that it took about 231 days for scientists to recognize Uranus as a planet and not a comet—the same number in years that will elapse between 1781 and 2012. Thus, in 1781, the solar system doubled in size, and a new way of marking time cycles became possible through the 84-year cycle that Uranus takes to revolve around the Sun.

The planet Uranus is associated with the sign of Aquarius— the Water Bearer. From an esoteric perspective, Uranus is connected with the seventh ray, one of seven energetic influences that powerfully shape human evolution. The seventh ray is the harbinger of the new age and will increasingly make its presence felt in human affairs. The seventh ray brings the impulse toward greater group cooperation, the building of new forms that allow Spirit to work more deeply on the material plane, and the power to understand and implement spiritual laws.[2] Great scientific progress in many areas of exploration, including the areas of spiritual healing, alternative medicine, and new energy technologies, are connected with the subtle vibratory influence of this ray. Astrologically speaking, Uranus is connected with the sudden appearance of transformative impulses that change the established order. Thus, it corresponds to new inventions, new social impulses, and new spiritual understanding. Uranus, the change-bringer, connotes a definite break from the past social order and established norms. Under the influence of Uranus, there will be more widespread recognition of these new discoveries by the year 2012.

WHEN DOES THE AQUARIAN AGE BEGIN?

The precession of the equinoxes, caused by Earth's very slow wobbling on its axis, leads to a shift of the position of the solstice and equinox points with reference to the fixed stars. This shift through the zodiac is 1 degree every 71.5 years. As a result, the spring equinox shifts from one tropical sign of the zodiac, spanning 30 degrees, to the previous sign about every 2,160 years. In our current cycle, at every spring equinox (when the tropical zodiacal sign Aries begins), the Sun can be seen as standing in the background of the constellation of the Fishes (Pisces). Currently we are completing the Age of Pisces, which began around two millennia ago. Depending on which reference point is used, in the future, astronomers will observe the spring equinox occurring in the background of the stars of the constellation Aquarius. At that time, the Aquarian Age will be said to have begun.

Yet to astronomers, zodiacal constellations can span larger or smaller regions of space and not simply cover equal areas spanning 30 degrees. In fact, stars that compose each sign of the zodiac are themselves moving constantly over large epochs of time. Thus, the imaginary outline that makes up the mythological figure of each constellation of the zodiac can have a different geometry and name when viewed from different cultural perspectives. As a result, it is difficult to know precisely where each sign of the zodiac starts or ends. Although there is a general consensus that we are at the end of the Age of Pisces, the Fishes, the actual year when the Aquarian Age officially begins continues to be controversial.

According to the Tibetan teacher Djwhal Khul, who dictated his teachings through the late Alice Ann Bailey, the Aquarian Age will begin in 2117. If we use this date in relationship to studying cycles of American history and future world events,[3] some interesting correlations with the year 2012 begin to emerge. As shown in Figure 17, the 336 years between 1781, when Uranus was discovered, and 2117, when the Aquarian Age is said to begin, span exactly four Uranus cycles of 84 years, or 16 subcycles lasting 21 years. The number 16 has traditionally been associated with the number of petals in the throat chakra, symbolic of humanity's creative self-expression and the power of the word.[4]

By 2012, 11 seasons of Uranus (lasting 21 years each) will have been completed since 1781. Only five more subcycles, spanning a time period of 105 years total, will remain until the Age of Aquarius begins. In other words, humanity will have the opportunity to collectively prepare itself for the age of universal brotherhood/sisterhood for over a century.

The cycles of the planet Venus also played an important role in the Mayan calendar, as it was used in their divination. The rhythms of Venus, which, as seen from Earth, appears to make loops along the zodiac during its inferior conjunctions with the Sun, recur with regularity. Venus-Sun conjunctions create a perfect pentagram in the zodiac during an exact eight-year period.

Much rarer are Venus's transits over the physical disk of the Sun, as seen from Earth. They recur four times in a 243-year rhythm made up of a 130-year period and followed by another 113-year period. These solar transits always occur in pairs,

FIGURE 17: 84-year Uranus cycles.

eight years apart. The last one occurred on June 8, 2004, and was watched on the Internet by thousands of people worldwide. Another passage of Venus in front of the disk of the Sun will occur on June 6, 2012. There will not be a similar Venus transit over the Sun's disk until December 11, 2117, the year the Age of Aquarius begins. That future passage will be followed by another paired event eight years later on December 8, 2125.

These passages of the planet Venus over the Sun bring in new cultural impulses. Rudolf Steiner, the founder of anthroposophy, said these events will bring about great changes. If we look back to the events close to 243 years ago, in June of 1761 and 1769, we see that great expeditions were sent forth from European countries to different parts of the globe to witness Venus's transits over the Sun from different locations on Earth. By comparing precise time measurements of these events, cooperating scientists and explorers were able to calculate the distance from the Earth to the Sun for the first time. This distance measure in space is called the astronomical unit (AU).

After 1769, the scientific determination of the AU gave humanity a yardstick by which to determine its place in the universe. Over time, the AU also changed humanity's perception of space. The passage of Venus in front of the Sun in 2012 will also impact humanity and is likely to accelerate its spiritual evolution toward greater collaboration and peaceful coexistence. The anticipated scientific developments associated with these events are discussed in greater depth at the end of this essay.

THE IMPORTANCE OF 33-YEAR CYCLES

Another important cycle relevant to 2012 spans 33 years and involves the balance between the solar and lunar calendars. A tropical solar year lasts 365 days, 5 hours, 48 minutes, and 45 seconds. The solar year is about 11 days longer than a lunar year (12 lunation cycles or synodic months) of 354 days, 8 hours, and 53 minutes. As a result, 33 solar years are equivalent to 34 lunar years as the cycles of the Sun, moon, and Earth come into a balanced rhythm once again. Scientists have also found that the Sun has an 11-year rhythm of solar activity marked by a higher number of sunspots. Sunspots are powerful magnetic braids in the solar corona that can also be accompanied by solar flares and powerful coronal mass ejections. A 33-year cycle, therefore, encompasses three 11-year cycles of sunspot activity. Sunspots, in turn, influence our cosmic weather, the water level of some inland lakes, the growth of trees as shown in their tree rings, and the deposit of minerals like clay and calcium in our soil.

Rudolf Steiner talked about the importance of the 33-year rhythm in this way: "This rhythm deals with the effect that each individual person has on their social sphere. What one does, including one's thoughts, affects the behavior between

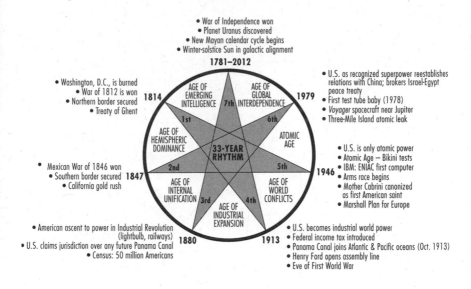

- War of Independence won
- Planet Uranus discovered
- New Mayan calendar cycle begins
- Winter-solstice Sun in galactic alignment

1781–2012

- Washington, D.C., is burned
- War of 1812 is won
- Northern border secured
- Treaty of Ghent

1814

AGE OF EMERGING INTELLIGENCE — 1st

AGE OF GLOBAL INTERDEPENDENCE — 7th

1979

- U.S. as recognized superpower reestablishes relations with China; brokers Israel-Egypt peace treaty
- First test tube baby (1978)
- *Voyager* spacecraft near Jupiter
- Three-Mile Island atomic leak

AGE OF HEMISPHERIC DOMINANCE

33-YEAR RHYTHM

ATOMIC AGE — 6th

- Mexican War of 1846 won
- Southern border secured
- California gold rush

1847

AGE OF HEMISPHERIC DOMINANCE — 2nd

ATOMIC AGE — 5th

1946

- U.S. is only atomic power
- Atomic Age — Bikini tests
- IBM: ENIAC first computer
- Arms race begins
- Mother Cabrini canonized as first American saint
- Marshall Plan for Europe

AGE OF INTERNAL UNIFICATION — 3rd

AGE OF INDUSTRIAL EXPANSION

AGE OF WORLD CONFLICTS — 4th

- American ascent to power in Industrial Revolution (lightbulb, railways)
- U.S. claims jurisdiction over any future Panama Canal
- Census: 50 million Americans

1880

1913

- U.S. becomes industrial world power
- Federal income tax introduced
- Panama Canal joins Atlantic & Pacific oceans (Oct. 1913)
- Henry Ford opens assembly line
- Eve of First World War

FIGURE 18: American history in 33-year cycles.

people within the larger social sphere 33 years later."[5] If one views the creative impulse of each human generation as maturing over a 33-year period, then the social impulses and ideas of each generation are transformed into a living act of will and collective purpose 33 years later. Seen in this way, each century has the social initiatives of many generations simultaneously active within it.

In Figure 18, one might also recognize a 33-year rhythm active in American history. By 2012, there will have been seven 33-year cycles since 1781, the year the United States won its independence and the planet Uranus was discovered. The first 33-year cycle ended in 1814, when the Treaty of Ghent was signed, effectively ending the War of 1812, which had also been called the Second War of Independence from the British. Thirty-three years later, in 1847, General Zachary Taylor, who would later become the twelfth president of the United States, was active during the Mexican-American War of 1846 that led to the permanent establishment of the southern U.S. border. This war also led to the later annexation of California.

As the third 33-year cycle completed in 1880, the United States had become a powerful industrialized nation and was completing its second transcontinental railway system as an artery of commerce. This would allow the nation to be able to become a world power in the twentieth century. The fourth 33-year cycle ended in 1913, on the eve of World War I. On October 10, President Woodrow Wilson pushed a button in Washington that ignited eight tons of dynamite 4,000 miles away to open the last segment of the Panama Canal, thereby joining the waters of the Atlantic and Pacific oceans in Central America. The year 1913 symbolized the ending of a cycle in the Age of Industrial Expansion, of which Henry Ford's automobile assembly line became an international symbol.

Between 1913 and 1946, the United States was involved in two world wars in an Age of World Conflicts. In 1945, just as the fifth 33-year cycle was completing, the United States initiated humanity into the Atomic Age with the explosion of the first nuclear bomb at the Trinity test site, located on the Alamogordo Bombing and Gunnery Range, near Socorro, New Mexico. This was followed by two further unfortunate atomic bombs that caused great death, destruction, and human suffering when they were dropped on the Japanese cities of Hiroshima and Nagasaki. These events led to the ending of World War II and the beginning of the sixth 33-year cycle that began in 1946. That cycle was characterized by the nuclear weapons race, the Cold War, and military policies of mutual assured destruction (MAD). By 1979, the United States had recognized the cost of being a nuclear superpower and experienced painful lessons of war in Korea, Vietnam, and many other clandestine conflicts.

Computers, the semiconductor industry, and the communications revolution were all being invented as the Information Age was in its ascendancy.

From 1979 to the present, the world has undergone a fundamental shift. Humanity has become increasingly interconnected through both economic globalization and ecological crises. Acting locally now influences everyone globally. One might say that we now live in an Age of Global Interdependence. Our patterns of war, trade hegemony, exploitation of underprivileged nations, destruction of our planetary ecosystems, and ocean and atmospheric weather changes due to widespread industrial policies and energy usage cannot continue without endangering our very survival. As this seventh 33-year cycle comes to an end in 2012, the United States is being invited to recognize both the global emergency as well as the new global emergence. The natural social cycles of 33 years are clearly apparent in American history as well as in world events. It is hoped that the United States will choose to become the "world disciple" and demonstrate moral leadership and inspired action for a new era in human relations.

A new blueprint for American leadership and world service is being invoked in 2012. Things are bound to change drastically in the twenty-first century, as China and India are becoming increasingly valuable economic partners along with our European and Japanese partners. The eighth 33-year cycle in U.S. history will complete in 2045. This is exactly 100 years after the United States first became internationally recognized as a global compassionate leader by creating the Marshall Plan to aid the rebuilding of Europe after the Second World War. The ninth and tenth 33-year cycles will be completed in 2078 and 2111, respectively. Six years later, the Aquarian Age is said to begin in 2117.

There are even longer cycles, spanning 26,000 years, that are now bringing changes into human consciousness. They involve the alignment of our Earth and solar system with the galactic coordinates of the Milky Way, a galaxy comprising an estimated 40 billion stars. The existence of these long cycles was known to ancient cultures. Thousands of years ago, Vedic, Egyptian, and, later, Mayan astronomers all understood the effect of this alignment on human consciousness.

A brief description of the various astronomical coordinate systems will help clarify the spatial alignment relevant to these longer cycles. The plane of the ecliptic is defined by the movement of the Earth around the Sun. Most of the planets in our solar system orbit within a few degrees above or below this ecliptic plane. In this plane, we also find the signs of the zodiac. In contrast, the galactic plane is defined by the various arms of the Milky Way galaxy, which collectively form a flattened disc with many spiral arms. The ecliptic plane is inclined to the galactic plane at an angle of about 61 degrees.

At this important moment in time, our winter-solstice Sun is aligned exactly with the equator of our galaxy. This has not happened for 26,000 years. Since the disk of the Sun, seen from the Earth, spans a diameter of one-half of 1 degree in width, we are now in a 36-year period during which the Sun's disk at the winter solstice makes this unusual contact with our galactic equator. This 36-year period is said to have begun in 1980 and will end around 2016.

In the past decade, astronomers have penetrated into the mysteries at the center of our Milky Way galaxy. The galactic center appears to be located near the astronomical region designated as Sagittarius A, located at 26 degrees 45 minutes of Sagittarius in zodiacal longitude.[6] Seen from Earth, the Sun aligns with this exact point just three days prior to the winter solstice, and in the future, this will occur ever closer in time to this point of galactic attraction.

The galactic center appears to be around 26,000 light-years from Earth. Scientists use light, traveling at 300 million meters per second, as a yardstick for distance. One light-year is a distance equivalent to about 5.88 trillion miles, making the galactic center about 153 quadrillion miles distant from Earth. As a result, visible light, X-rays, gamma rays, or radio emissions that reach us today from the galactic center took 26,000 years to arrive on Earth.

Since a complete precession cycle of the equinoxes also takes around 26,000 years, we might recognize a most important synchronicity here. The light that we are now seeing from the center of our Milky Way galaxy originated the last time the winter solstice Sun aligned with the galactic center, exactly one 26,000-year cycle ago. This might suggest that humanity is about to undergo a "quantum leap" in its conscious evolution.

Scientists have been working tirelessly to penetrate the "heart of the galaxy." Researchers have found that in the central "heart" of our galaxy resides a giant black hole, some 2.5 million times as heavy as our Sun. This quiescent center is the gravitational still-point around which billions of stars and solar systems slowly rotate. Gravitational waves, as well as electromagnetic radiation consisting of high-energy cosmic rays, gamma rays, and X-rays, are emitted constantly from this galactic center region. These powerful emanations are radiated in all directions, ultimately reaching our Sun and the Earth. Every year, near the winter solstice, our Sun is uniquely placed to directly absorb the greatest amount of the radiation from the billions of stars surrounding the galactic center's black hole. This living starlight radiates its life-giving energy and information to our awakening Earth.

In just the last four years, scientists have, for the first time, had a clearer view into the mysterious galactic center of our own Milky Way. Shrouds of galactic dust make this area invisible to our optical telescopes, and we can only see faint glimmerings of light from this region. It has always been invisible to our naked eye, much like the unconscious has been "invisible" to our conscious awareness. However, with the help of new scientific instruments such as NASA's orbiting Chandra X-ray Observatory, as well as arrays of radio telescopes and near-infrared light imaging technologies, scientists for the first time have glimpsed deeply into the "city of light" at the center of our galaxy. Thus, in our time, we are witnessing an alignment of our zodiacal precessional cycle, spanning about 26,000 years, with the exact heart-center in the plane of our galaxy.

WHAT CHANGES CAN WE EXPECT AFTER 2012?

There will be many great breakthroughs in science and technology, medicine and art, culture, and religious understanding. Today's cutting-edge inventions, including quantum computing, new nanotechnology materials engineering, artificial intelligence, and genetic engineering, will be greatly expanded. What is on the horizon includes elemental transmutation, unlimited energy from the quantum vacuum states, silent transportation systems based on levitation principles, epigenetic engineering, light and sound healing through energy medicine modalities, and new theories of electrodynamics that collectively presage a massive shift in human creativity in the twenty-first century.

In the scientific arena, a new understanding of electrodynamics is beginning to emerge. Around 100 years ago, Nicola Tesla, the inventor of modern alternating current technology, described longitudinal waves, or "Tesla radiation," that were markedly different from the nineteenth-century electromagnetic theories of James Clerk Maxwell, the theoretical physicist who developed the first field theory of modern physics. While Maxwell spoke of transverse electromagnetic waves moving at the speed of light, Tesla spoke of scalar or longitudinal waves that work through resonance processes and have great relevance for medicine and healing. This scalar wave technology is already at work in cell phones, although this is not generally understood in the electrical engineering community.[7]

Russian and German scientists have also discovered that we live in an ordered, interconnected universe where all matter is organized according to a fractal mathematical order. This new understanding is called global scaling. These discoveries describe the harmonic structure of logarithmic space, explain the nature of our universal constants, and define the nonlinear, but fractal, hyperbolic scales that organize the universe.[8] Global scaling will revolutionize communication technologies, create more environmentally friendly building designs, and impact biotechnology and medicine as well as antigravity research.

Light technologies will rapidly advance to revolutionize human health and well-being. The body communicates at the cellular level through coherent light called biophotons.[9] This radiation from living tissues was originally discovered by Dr. Alexander Gurwitsch in Russia in the 1920s, who called it mitogenic radiation. Today, using biophoton research, scientists in many parts of the world are studying the effect of agricultural methods, food-processing technologies, and water treatment systems on cellular health.

Other scientists have found that modulated coherent light from lasers and lighted crystals have a powerful healing effect on our cells and organs. The healing modalities of the future will utilize coherent light and modulated light to not only affect cells, tissues, and organs, but also to modulate the body's subtle energetic system—the human chakra system—and directly influence the healing process. Low-intensity light and magnetic emanations have already been measured from the hands and chakras of natural healers by sensitive photomultiplier tubes and magnetometers. This will create the basis for a future science of magnetic healing, through the laying on of hands and radiatory healing, also called distant healing, by which many people can be healed.

In addition, high-energy light will also be harnessed for new ways of energy production from microscopic bubbles created in water by resonant-sound technologies. Under certain circumstances, harmonic sound interacting with water is capable of creating intense light similar to that found in nuclear fusion reactions. This new technology may be capable of becoming a source of unlimited, pollution-free energy. Current scientific research into this phenomenon, called sonoluminescence, is now being actively pursued in laboratories in many countries.[10] By using inert gases, such as xenon, krypton, argon, neon, and helium, the process can be greatly intensified.

These noble gases are also being used in low-pressure tubes to generate "plasma waves" for healing many diseases, including cancer. These gases, when stimulated by modulated electromagnetic radio-frequency waves, have been recognized for their ability to assist in healing imbalances in our bodies and are going to be a part of the new field of energy medicine healing.

In the twenty-first century, through the science of epigenomics, environmental conditions will increasingly be recognized for their capacity to affect genetic expression. Scientists now know that the human genome is essentially controlled by special regulatory proteins that turn genes on and off according to many environmental influences. Much of the noncoding DNA, currently called "junk DNA," will also be utilized to modulate genetic expression. New electromagnetic modalities, including modulated laser light, will rapidly induce healing of genetic imbalances by affecting the transfer of genetic information. This has already been demonstrated by Russian scientists.

Ultimately, it will be recognized that not only physical factors, but also human thinking and emotions, can affect genetic expression and disease manifestation.[11] As society recognizes that thoughts and emotions are powerful modulators of our biological codes, new educational systems will arise to teach people how to control their emotions and thoughts through self-regulated biofeedback. This will rapidly advance the current field of mind-body medicine and lead to a new collective awareness of the effect that our feelings and belief systems have on our health. The current science of psychoneuroimmunology is already laying the scientific basis for this new medical approach.

Scientists have recently discovered that we are uniquely different from one another because we have multiple copies of the various genes and not just differences in the base pair sequences that make up the human genome, or "book of life." Until now, it was assumed that the human genome was largely the same for everyone, except for a few "spelling" differences in some of its genetic "words." The latest findings, however, suggest that the book of life contains entire sentences, paragraphs, or even whole pages that are repeated any number of times. These new findings mean that instead of humanity being 99.9 percent identical, as previously believed, we are at least 10 times more different from one another than we once thought. This could explain why some people are prone to serious diseases.

As we move away from the current mechanistic approach to genetics to understanding that the genome is fluid in nature and is capable of being greatly influenced by many environmental factors, personal nutrition, and even emotional states and mental beliefs, there will be greater personal responsibility for our lifelong health creation and maintenance. It may also help to explain new processes that could be involved in spontaneous healings, as well as how spiritual practices may positively affect our health.

By 2012, our understanding of harnessing energy from the universe will change. Scientists have already demonstrated that 96 percent of the universe is made up of an unknown dark energy (73 percent) and a mysterious dark matter (23 percent). This leaves only 4 percent of the known universe capable of being directly probed by our current scientific instruments. According to astrophysicists, we live immersed in a virtually infinite supply of energy called the quantum vacuum state (or zero-point energy domain). The matter/energy density has been calculated to be as high as 10^{94} grams per cubic centimeter of space.[12] When humanity learns to access this virtual particle flux present in the vacuum, humanity will never go back to inefficient fossil fuels, radioactive nuclear energy, or other polluting technologies that contribute to global warming. Wars fought for control of these polluting hydrocarbon resources will become relics of the past, heralding a new era in human evolution.

New forms of atomic lattice nanotechnology engineering will utilize scientific principles, such as the Casimir effect, to allow humanity to create novel materials that can, in turn, create a resonance energy transfer out of this sea of unlimited energy that surrounds us. Other discoveries leading to the creation of a hydrogen energy economy are now being made. Researchers have discovered previously unrecognized electron-low energy states, called hydrinos, within the hydrogen atom. The energy released from this process is hundreds of times in excess of the energy required to start it because it draws energy out of the sea around us, also called the quantum plenum. The primary fuel is hydrogen gas, which can be created inexpensively from water via electrolysis. This work is based on a new theory, developed by Dr. Randell Mills, called classical quantum mechanics—an apparent unified field theory that unites quantum physics with Maxwellian electrodynamics and Einstein's theories of special and general relativity. This technology promises to create a source of clean, cheap energy from the untapped energy states within the hydrogen atom derived from water.[13]

By using these forms of energy production and conversion, the forces of levitation will be harnessed. New transportation systems that literally float through the air will be invented. These transportation systems will be quiet and clean and will use the antigravitational forces inherent in the new theories of expanded electrodynamics. Science will discover a new form of magnetoelectrism,[14] in contrast to the present one-sided electromagnetic phenomenon. Free energy from overunity magnetic motors, having already been invented on several continents, will also be part of this twenty-first century technology.

The limitations of the human mind will be overcome, and unbounded creativity promises to become humanity's destiny. According to Huston Smith, each brain is already capable of a greater number of associations than the number of atoms in the universe. Moreover, there is strong evidence that the holographic universe acts as an interconnected mind to which each human being can have direct access. Humanity is already capable of remote viewing, telepathy, and clairvoyant vision. The reality of humanity's interconnected thought field has already been rigorously demonstrated by psychophysiological researchers and found to be based on quantum physical principles called entanglement.[15]

New capabilities of clairvoyant sight will become more commonplace in future generations. Some signs that this is developing can already be seen in the demonstrated abilities of some unusual children, who have been given names such as Indigo, Dolphin, or Crystal children. These children are demonstrating advanced mental and emotional capabilities that will become commonplace within all of humanity. There will also be breakthroughs in interspecies communication technology, allowing humans to have dialogues with intelligent creatures such as whales and dolphins. This will aid humanity in becoming better environmental stewards of the planet.

By 2012, human consciousness research will also be expanded to include scientific investigations into the instrumental transcommunication process with deceased persons. Foundational work by many investigators on several continents, using computers, tape players, video recorders, and radios, has already led to the recording of voices and images from known persons who have passed over. These developments will lead to greater acceptance of the laws of reincarnation that are already part of many world religions. These principles are also not new in the West, for it is a historical fact that the concept of reincarnation existed within Christianity until condemned at the fifth Ecumenical Council held in Constantinople in May A.D. 553. The recognition that human beings reincarnate repeatedly will powerfully impact humanity's modes of behavior and lead to greater awareness of the impact that our thoughts, emotions, and deeds have on our future lives.

There will also be widespread recognition that spiritual intelligences of many qualities have always been present in our midst. These nonphysical beings—whether we call them *devas*, angels, celestial intelligences, or extraplanetary influences—will establish more direct contact with many sensitives. There is a long history of unusual experiences and illuminating contacts in every human culture. The unusual appearance of unexplainable large plants and vegetables grown in the Findhorn Gardens in Scotland in the 1970s is one example of an experiment in human-angelic cooperation.[16] These celestial intelligences are glad to assist humanity during this time of global transition into a new era. This will ultimately lead to a universal spirituality; one not tied to any specific religious tradition, but rather a synthesis of the highest intentions and ideals within all the world's religions.

One of the greatest hindrances to human advancement at this time is the lack of willingness to change the centralized economic power structures and financial controls that are now vested with relatively few, and disproportionately powerful, organizations and corporations. New models of economies of sharing and mutual empowerment will need to arise. This has already begun with certain forward-looking, globally oriented business practices. Starting in 2012, there will be greater international affirmation that humanity has become increasingly globally interdependent. What is now needed is the creation of new economic measures and cooperative governance processes that support the healthy development of global exchange and sharing. Ultimately, the principles of love, sharing, and mutual goodwill must begin to govern human relationships, whether personal, national, or international.

Beginning in 2012, humanity will systematically learn to leave the "matrix of scarcity" that has kept humanity enslaved in poverty, due to lack of sharing and erroneous financial beliefs, including a debt-based monetary system. One of the triggers to a shift in economic consciousness may be the widespread recognition by many U.S. citizens that the United States has incurred the largest global debt of any country on the planet. Due to the size of this debt, amounting to nearly 70 trillion dollars and equivalent to five times the country's current gross national product, some economists have suggested that the United States could be bankrupt.[17] The U.S. economy is being sustained by the savings from around the world. At present, these international monies already amount to one-third of annual U.S. domestic investment. As all world economies become interdependent, a global financial reorientation will undoubtedly take place. Many new breakthroughs in economic and scientific research will come from countries that are now only in their ascendancy.

Our collective future is full of challenges and promises that will require local, national, and global awareness and action. Humanity is at a point that it must make a decision as to what future it seeks to create. May the year 2012 be seen as an important nexus in time that stands as a doorway to global peaceful cooperation and sharing on this interconnected and increasingly interdependent planet.

2012

SOCIALLY RESPONSIBLE BUSINESS AND NONADVERSARIAL POLITICS

CORINNE McLAUGHLIN

It is often said that the 2012 changes will leave nothing in our world untouched. Many talk about how 2012 will affect spirituality and consciousness, but will our established patterns of business and politics be transformed as well? Is there a way to use spirituality and consciousness to influence business and political trends? In the following essay, Corinne McLaughlin, author of Spiritual Politics *and* The Builders of the Dawn, *answers these questions and tells us how these arenas are already shifting and what we can expect in the next decade.*

Are you stuck in fear and dread about looming disasters culminating in 2012? If so, I'd like to help shift the way you look at the world, especially business and politics. Why not imagine a better world in the future and live this vision moment to moment in the choices you make and the work that you do? In this way, you can help create it.

We, as humanity, have a choice: do we want to see the physical and emotional suffering of the old world increase, both personally and collectively, or do we want to experience more of the beauty of the new world? As members of the human family, each and every one of us is facing a great soul choice: which path will we take in the coming years of upheaval and change?

The choice is between holding on to old self-centered patterns of thinking or opening ourselves to new values and more peaceful and sustainable ways of living. One choice leads to escalating crises and chaos—learning the hard way through increased physical and emotional suffering brought on by selfishness and greed. The other choice offers the evolutionary opportunity to be more compassionate and caring toward others, and so build a new civilization. Although there is a higher nondual reality beyond this seeming duality of new and old worlds, right now, at our human level, we must make choices and act.

Did you know that at this very moment a new world is emerging right through the cracks and crevices of the old world? It's alive, growing, and vibrant, in stark contrast to the old world, which is running on fear, anger, and greed. By 2012, this new world, born out of the creative minds and compassionate hearts of self-empowered visionaries everywhere, will be even more visible and influential, affecting every aspect of life.

These practical visionaries, whom I call "the builders of the dawn," are found in all fields of life and in all nations. They're bringing new ideas and solutions to our problems—from war and terrorism to poverty and environmental destruction. These visionaries often seem like they are glowing from within, as if they are filled with light and passion.

It's easy to see the negative impact of business and politics in the world today—just pick up a newspaper or turn on the evening news. In case you've missed them, however, there are also many positive changes in business and politics going on around you. I've been researching this growing phenomena for some time, and over the years, I've met many of the founders and directors of these new projects.

I'd like to help connect the dots in business and politics so that you can see the bigger pattern emerging. The new world is growing steadily, and you can see proof of this if you know where and, more important, *how* to look. It's all a matter of perception. Although reports on these new approaches can be found in the mainstream media, few people are weaving them together to show the bigger picture of change now occurring. If we examine the positive trends in business and politics, it will give us great hope for the future. These new trends are more aligned with the direction of evolution—toward greater synthesis, interconnection, and creative intelligence.

One of the reasons for the growing impact of this new world by 2012 is simple: the generation that will be coming of age then is already more in tune vibrationally with it. And if they haven't yet discovered it, they soon will.

Some patterns of change are "continuous," meaning we can see certain trends and project them into the future; other patterns are "discontinuous," meaning they are based on chance or chaos. Because we live in a world of interlaced systems, where everything is connected to everything else, it's essential to notice the patterns and interconnections—and to be prepared for unexpected change as well.

SOCIALLY RESPONSIBLE BUSINESS

By 2012, multinational corporations will be larger and more powerful than they are today. They will affect every aspect of our lives, as the merging of giant companies continues and new markets are exploited around the world. While capitalism is extremely efficient at producing and distributing products and services, it has caused many damaging effects on both people and the environment because of its narrow pursuit of a single purpose: generating profit by any means necessary. To eliminate the damaging effects, the purpose of business needs to be expanded. Seeing money as the single bottom line will be a rarity in 2012. In a post-Enron world, values and ethics are an urgent concern. I've found that the hottest buzz today is the idea of a "triple bottom line"—a commitment to "people, planet, and profit." Concern for the welfare of employees and the environment is helping the bottom line, and recent studies have proven it.

Both external and internal factors are pressuring businesses to change. External forces, such as depleting oil supplies and strife in oil-producing countries, will continue to drive prices ever higher across the board, at the same time motivating companies to reduce their energy consumption and follow a more sustainable approach. Even major companies have begun to see the writing on the wall, and in their public relations campaigns, they compete to be seen as the most environmentally sustainable. Growing political pressure from citizens will lead to more government regulation of corporate excesses.

Responding to internal pressures, companies have also been transforming from the inside out. The movements for social responsibility and spirituality at work are growing dramatically and will be very significant by 2012. Social responsibility addresses the external effect of business on society, and spirituality addresses the personal, internal dimension.

For years, I've been investing in socially responsible businesses, leading trainings on spirituality at work, buying environmentally friendly products, and inviting visionary business leaders to speak at conferences I've organized. I've also been active in the international Spirit at Work organization and a fellow of the World Business Academy, which promotes corporate responsibility. All of this has greatly increased my optimism about the world in 2012.

Lifestyles of Health and Sustainability (LOHAS) reports there is now a $228.9 billion U.S. marketplace for goods and services focused on holistic health, the environment, social justice, personal development, and sustainable living. Significant financial investments are being redirected into these areas. Sociologist Paul Ray writes that approximately 30 percent of the adults in the United States, or 50 million people, are currently consumers of these goods and services. These people generally make conscientious purchasing and investing decisions based on their social and cultural values. They are the future of business and of progressive social, environmental, and economic change in the United States and elsewhere.

Each day, more and more businesses are helping create a better world by becoming more socially responsible—to their employees, their shareholders, their community, and the environment. Their financial success is turning heads. In her book *Megatrends 2010*, author Patricia Aburdene calls this "conscious capitalism" or "stakeholder capitalism" and names it as one of the emerging megatrends. She notes that socially responsible corporations tend to be well managed, and great management is the best way to predict superior financial performance.

The socially responsible movement provides concrete evidence that business, as the most powerful institution in world today, may be transforming from within. As World Business Academy cofounder Willis Harman remarked many years ago, "The dominant institution in any society needs to take responsibility for the whole, as the church did in the days of the Holy Roman Empire."

Business for Social Responsibility (BSR) is a San Francisco–based nonprofit founded in the 1990s, which has grown to encompass over 400 organizations, including about half of the Fortune 500 companies. BSR defines corporate social responsibility (CSR) as a "comprehensive set of policies, practices, and programs" that earn financial success in ways that "honor ethical values and respect people, communities, and the natural environment."

In the field of social responsibility, early pioneers that have developed well-respected brands and loyal costumers include The Body Shop, Ben and Jerry's, Stonyfield Farm, The Timberland Company, Patagonia, Tom's of Maine, and The Men's Wearhouse. These companies typically support community projects and good causes and find innovative ways to support their employees and protect the environment.

THE WORK/LIFE BALANCE

Internal pressures for change in the corporate sector arise when people realize that there's more to life and business than profits alone. There's been an increasing emphasis on what's called "work/life balance," as the work-addicted find a way "to get a life" and carve out time for other interests and needs outside of work. I've been part of a new wave of consultants and organizations that help individuals and companies find a way to achieve this balance. I love giving seminars on this theme, as happier, healthier employees boost profits. This movement is a major drive for change that will be far more dramatic by 2012.

People at all levels in the corporate hierarchy increasingly want to nourish their spirit and creativity. When employees are encouraged to express their creativity, the result is a more fulfilled and sustained workforce. Happy people work harder and are more likely to stay at their jobs. A study of business performance by the highly respected Wilson Learning Company found that 39 percent of the variability in corporate performance is attributable to the personal satisfaction of the staff.

Spirituality was cited as the second most important factor in personal happiness (after health) by a majority of Americans questioned in a *USA Weekend* poll. An astounding 47 percent said that spirituality was the most important element of their happiness. Increasing numbers of people think it's time to bring their spiritual values (but not necessarily their religion) into their workplace. Key spiritual values in a business context include integrity, honesty, fairness, accountability, quality, cooperation, intuition, trustworthiness, respect, and service.

To the surprise of many, this movement is beginning to transform corporate America from the inside out, and by 2012, it will have rippling effects on the economy and the social fabric. Growing numbers of businesspeople want their spirituality to be more than just faith and belief; they want it to be practical and applied. They want to bring their whole selves to work—body, mind, and spirit. Many businesspeople are finding that the bottom line can be strengthened by embodying their values. They can "do well by doing good."

Around the world, people increasingly want to bring a greater sense of meaning and purpose into their work life. They want their work to reflect their personal mission in life. Many companies are finding that the most effective way to bring spiritual values into the workplace is to clarify the company's vision and mission and to align it with a higher purpose and deeper commitment to serve both customers and the larger community.

HOW SOCIAL RESPONSIBILITY HELPS THE BOTTOM LINE

Are social responsibility and profitability mutually exclusive? Bringing ethics and values into the workplace can lead to increased productivity and profitability, as well as employee retention, customer loyalty, and brand reputation, according to a growing body of research. More employers are encouraging spirituality as a way to boost loyalty and enhance morale.

In their *Corporate Social and Financial Performance Report*, Mark Orlitsky of the University of Sydney and Frank Schmidt and Sara Rynes of the University of Iowa reviewed studies over the last 30 years and found a significant relationship between socially responsible business practices and financial performance that varied from "moderate" to "very positive."

Another meta-study, by the United Kingdom's Environment Agency and Innovest Strategic Value Advisors, called *Corporate Environmental Governance* reviewed 60 research studies and found that 85 percent demonstrated a positive correlation between environmental management and financial performance. The monetary rewards of this approach guarantee that it will grow dramatically in the coming years.

A study done at the University of Chicago by Professor Curtis Verschoor and published in *Management Accounting* found that companies with a defined corporate commitment to ethical principles do better financially than companies that don't make ethics a key management component. Public shaming of Nike's sweatshop factory conditions and slave wages paid to overseas workers led to a 27 percent drop in its earnings several years ago. Shocking disregard of ethics and subsequent scandals has led to financial disaster for companies like Enron, Arthur Andersen, WorldCom, Global Crossing, and many others.

Business Week reported that 95 percent of Americans reject the idea that a corporation's only purpose is to make money, and 39 percent of U.S. investors say they frequently check on business practices, values, and ethics before investing. The Trend Report found that 75 percent of consumers polled say they are likely to switch to brands associated with a good cause if price and quality are equal.

WHY SPIRITUALITY IS SO POPULAR

Researchers point to several key factors in the growth of spirituality at work, which will lead to a major impact by 2012. Corporate downsizing and greater demands on remaining workers has left them too tired and stressed to be creative. This is happening at the same time that globalization of markets requires more creativity from employees. To survive in the twenty-first century, organizations must offer a greater sense of meaning and purpose for their workforce.

In today's highly competitive environment, the most talented people seek organizations that reflect their inner values and provide opportunities for personal development and community service, not just bigger salaries. Unlike the marketplace economy of 20 years ago, today's information- and services-dominated economy requires instantaneous decision making and building better relationships with customers and employees.

Since most people today spend more time at work, there is less time available for religious activities after work. The *New York Times* recently reported that a growing number of companies are allowing employees to hold religion classes at work. This accommodates busy professionals who are pressed for time and afraid they have abandoned their faith. Many people are feeling more comfortable in the public expression of their faith.

Another factor in the popularity of spirituality at work is that there are more women in the workplace today, and women tend to focus on spiritual values more often than do men. The aging of the large baby boom generation is also a contributor, as boomers find that materialism no longer satisfies them, and they begin to fear their own mortality.

MEDITATION AND PRAYER AT WORK

Meditation classes (usually without religious content) are now held at many major corporations, such as Medtronic, Apple

Computer, Google, Yahoo!, McKinsey & Company, Hughes Aircraft, IBM, Cisco Systems, and Raytheon Company. Apple Computer's California offices have a dedicated meditation room, and employees are actually given a half-hour a day on company time to meditate, pray, or just sit in silence, as the company finds it improves employee productivity and creativity.

In the Washington offices of the Center for Visionary Leadership, we'd pause each day at noon for a quiet meditation and invite anyone who was visiting at the time to join us—surprisingly, most did. By 2012, we're likely to see more widespread embrace of meditation as an antidote for our stressful lives, while businesses note the financial benefits.

Medtronic, which sells medical equipment, pioneered a meditation center at their headquarters 20 years ago, and it remains open to all employees today. Medtronic founder, Bill George, says the purpose of business is "to contribute to a just, open, and sustainable society." He describes a "virtuous circle" whereby motivated, satisfied employees produce satisfied customers, which produce good financial results, which benefit the shareholders.

Many people use prayer at work for guidance in decision making, to prepare for difficult situations, when they are going through a tough time, or to give thanks for something good. ABC's *World News Tonight* reported that the American Stock Exchange has a Torah study group; the Boeing company has Christian, Jewish, and Muslim prayer groups; and Microsoft Corporation has an online prayer service.

Corporations are hiring chaplains to support their employees, because chaplains are good listeners and quick responders in crises and can serve people of any (or no) faith. Tyson Foods, for example, has 127 part-time chaplains at 76 sites; Coca-Cola Bottling Company has 25 chaplains, serving employees at 58 sites. Fast food companies such as Taco Bell and Pizza Hut hire chaplains from many faiths to minister to employees with problems and credit chaplains with reducing turnover rates by one-half.

ENVIRONMENTALLY SUSTAINABLE BUSINESSES

Environmentally "green" practices are becoming ever-more popular and stylish, as recent covers of *Newsweek* and *Time* magazines attest. With an emerging consensus among scientists and policymakers that global warming is real, sustainable business practices that protect the environment and reduce global warming will become widespread by 2012—especially as companies recognize that these practices also help the bottom line.

A Vanderbilt University analysis found that in 8 out of 10 cases, low-polluting companies financially outperformed their dirtier competitors. In recent years, over 300 multinational corporations have joined the United Nations Global Compact, pledging to support environmental protection, human rights, and higher labor standards. Companies not reporting significant progress in these directions each year are eliminated from the compact.

Founded in 1989, CERES, a coalition of investment funds, environmental organizations, and public interest groups, has been at the leading edge in pushing businesses toward environmental stewardship and greater transparency in public disclosure. The CERES Global Reporting Initiative is now the de facto international standard for corporate reporting on economic, social, and environmental performance.

Following the lead of small innovative companies, which have laid the foundations for years, many large multinational corporations are now making major changes. More than 2,000 pioneering San Francisco Bay Area firms are now certified as "green businesses" by the Alameda, California, county government and the Sustainable Business Alliance. California recently became the first state in the nation to cap greenhouse gas emissions, providing further motivation for businesses to become more sustainable.

Corporate giant Wal-Mart recently made a huge move into organic foods, eliminating chemical fertilizers, antibiotics, and so on. Amory Lovins, cofounder of the Rocky Mountain Institute and a world-respected pioneer in energy efficiency, is working closely with Wal-Mart to reduce greenhouse gases. The company also pledged to run entirely on renewable energy and produce zero net waste. It committed to doubling its huge truck fleet's fuel efficiency in 10 years—saving $300 million in fuel costs per year.

While Wal-Mart's track record of concern for the welfare of its employees or the communities where it builds is infamously dismal and lacking in social responsibility, its bottom-line calculation of the profitability of its environmental moves will motivate other companies in similar directions. Wal-Mart's potential influence on its worldwide supply chain could be far greater than that of the U.S. government.

There are three signs that demonstrate a company's authentic conversion to more enlightened practices: (1) publicly announced specific goals and timetables, (2) buy-in at every level of the company, and (3) transparent reporting. Wal-Mart will be closely watched, as will other companies in the years ahead.

Interface Carpets, the world's largest commercial carpeting manufacturer, trained 8,000 employees in environmental sustainability, with the goal of reducing pollution to 0 percent. Instead of buying a carpet from Interface, you now rent a carpet, and when it wears out, you bring it back to be recycled and are given a new recycled one. The company has saved $185 million on waste reduction efforts alone.

Whole Foods, where I've shopped weekly for more than 10 years, is the world's leading natural and organic foods supermarket. It recently made the largest renewable energy purchase anywhere to offset 100 percent of its electricity use in all 180 stores, and it is the only Fortune 500 company to do so. It is purchasing more than 458,000 megawatt-hours of renewable energy credits from wind farms. This will have the same environmental impact as taking 60,000 cars off the road or planting 90,000 acres of trees. For nine consecutive years, Whole Foods was ranked by *Fortune* magazine as one of the 100 Best Companies to Work For.

Ten years ago, I had the honor of interviewing Dr. Karl-Hendrik Robert, founder of The Natural Step in Sweden. Dr. Robert developed a framework based in natural science that enables businesses and other organizations to integrate environmental considerations into strategic decisions and daily operations. In Sweden, The Natural Step was extremely effective in building a consensus among scientists and businesses on the urgent need for environmental sustainability, so that businesses recognized the competitive advantage of sustainability and actually lobbied the government for green taxes—to tax-laggard companies. Author Paul Hawken brought The Natural Step to the United States more than 10 years ago, and it is working with many major U.S. companies.

Sweden-based Electrolux adopted The Natural Step framework and phased out chlorofluorocarbons in its refrigeration systems. It reduced water use in its washing machines from 45 to 12 gallons and substituted canola oil for petroleum-based oil in its chainsaws.

SOCIAL INVESTMENT
More and more people want to invest in companies that embody values they care about—social, environmental, ethical,

etc.—and this trend will grow exponentially by 2012. In addition to this bottom-up movement of individual investors, when corporations like Enron, WorldCom, and Tyco suffered heavy damage due to ethics scandals, even Wall Street's institutional investors suddenly saw the light. Pioneering Calvert Social Investment Funds notes that today's social issues have a way of becoming tomorrow's economic problems; so it says social responsibility makes good business sense.

In 1994, socially responsible investing (SRI) was already a healthy $40 billion market, but by 2005, it had grown to a $2.29 trillion industry, which includes the three social investing strategies of screening, shareholder advocacy, and community investing. For the last 10 years, SRI has grown 40 percent faster than the overall fund universe. So by 2012, it could be huge.

There are three major SRI strategies, the first of which is screening. Socially responsible mutual funds subject stocks to a set of "screens," or criteria, asking, for example, Does the company pollute the environment, violate fair labor practices, promote women and minorities, display integrity in advertising? Many socially responsible investment funds avoid companies that produce firearms, nuclear power, tobacco, and alcohol.

My husband, Gordon Davidson, was the first executive director of the Social Investment Forum, and he would often field questions from major media, such as the *Wall Street Journal*, who were skeptical until they saw the stellar financial performance of SRI funds.

Shareholder advocacy is another powerful SRI strategy, where shareholders have pressured major corporations such as McDonald's and JCPenney to be more socially responsible through shareholder resolutions and divestment campaigns. The Shareholder Action Network successfully lobbied the Securities and Exchange Commission to help investors obtain full voting disclosure from their mutual funds, and they developed a Web site identifying mutual funds with the least and most responsible voting records. This growing shareholder movement could be a major force in the market by 2012.

Community investing is a third strategy that encourages people to invest in valuable local projects that might not qualify for funding. Chicago's South Shore Bank buys abandoned, deteriorating buildings and rehabs them, creating good jobs and safe neighborhoods and driving out drug dealers. Green banks, such as Triodos in the Netherlands, invest in organic farms, sustainable energy, and microfinancing worldwide. Grameen Bank in Bangladesh, FINCA in South America, and Women's World Banking are several pioneering examples of microenterprise financing, where small loans are made to the poorest of the poor, with an astonishing 99 percent payback rate.

NONADVERSARIAL POLITICS

Business is not the only field that is transforming itself both from within and without. Politics is being transformed in many significant ways that could have a huge impact by 2012. Like the "spirit in business" movement, a new "spiritual politics" is emerging—on the left as well as the right (with books such as *God's Politics, The Left Hand of God*, and *Spiritual Politics*, which I coauthored). Major conferences organized by the Tikkun Network of Spiritual Progressives, Sojourners, and the Center for Visionary Leadership have drawn thousands of people and much media attention. Speakers have highlighted poverty, the environment, and social justice as spiritual issues, rather than highlighting conservative issues such as abortion and stem cell research.

The growing spiritual activism of the progressive left in the United States is countering a decade of intense activism by the fundamentalist right. For the first time, many Democrats openly discussed their faith and spiritual life and won extra votes in the 2006 elections because of it. According to an exit poll by Zogby International, poverty and economic justice topped the list of the "most urgent moral problem in American culture," rather than issues favored by conservatives, such as abortion or same-sex marriage. This focus on the spiritual dimension will be changing the face of politics in the next few years, as people increasingly apply their spiritual values in the political arena and confront their differences with one another.

Another significant trend is the dramatically increased power of the "independent sector" or "civil society," which is the biggest movement in the history of humanity, although it is largely unrecognized and flies under the media radar. It has no center, codified beliefs, or charismatic leader. Its power is based on ideas, rather than on force. By 2012, it will be seen as the new superpower in the world. Because government hasn't been very effective at addressing major problems such as poverty, war, violence, terrorism, and environmental pollution, nonprofit organizations and citizen activist groups worldwide have taken up the challenge.

In his book *Blessed Unrest*, visionary activist and entrepreneur Paul Hawken has documented an astounding one million or more of these groups worldwide. In a recent talk to the Bioneers conference, he called this movement "humanity's immune response to resist and heal political disease, economic infection, and ecological corruption caused by ideologies. . . . It is everywhere. There is no center. There's no one spokesperson. It's in every country and city on Earth, within every tribe, every race, every culture, and every ethnic group."

Another arena that is transforming politics is the powerful new Internet "webocracy," which has enabled more direct democracy and has changed the political landscape dramatically in just a few, very short years. Community sites that draw millions of fans suddenly have major political influence, and this will have unkown ripple effects by 2012 and beyond.

Internet activism, exemplified by sites such as MoveOn.org and others of all political stripes, are affecting political races and issues. Web 2.0 media sites such as Digg.com and political blogs such as DailyKos.com are doing end runs around mainstream media and are becoming so popular that even mainstream politicians court them.

Voter drives on widely celebrated youth sites such as MySpace .com have registered thousands of new voters. Mobilevoter.org helps organizations set up technology, so people can text in their voter registration from their mobile phones.

Video-sharing sites such as YouTube.com enable people to share political video clips with millions of users. Online booksellers such as Amazon.com make every political book and exposé easily accessible to all. Search engines like Google.com provide more transparency in politics, as anyone can quickly access information on the Web.

Another major factor that will impact American-politics in the future is the fate of voting machines, as so much evidence has recently surfaced putting in question their accuracy and ability to be tamper-proof. If these challenges to voter protection are not resolved soon, they will undermine confidence in democratic institutions and lead to huge unrest by 2012.

CONFLICT TRANSFORMATION TECHNIQUES
The political process itself is transforming, as new mediation techniques effectively resolve conflicts and multistakeholder citizen dialogues make better policy decisions. These new approaches are emerging in many places around the world,

even amid the increasing polarization of American politics and the so-called "Red-Blue Divide." Participants on either side of a conflict do not have to give up their deeply held values; rather, they must find common interests upon which they can act together.

A nonadversarial approach, which seeks higher common ground on polarized issues, could become even more widespread and influential by 2012. This new politics is emerging most clearly in the global civil society and among the new "social entrepreneurs," but we also catch glimpses of it in state and local governments, and even occasionally in national governments.

Conflict resolution techniques reduce violence by helping people listen more deeply to voices on all sides of an issue, even in the midst of generations of ethnic strife. These techniques help people recognize their common humanity. Unity is needed before there can be lasting peace in the world, and peace is needed before there can be shared abundance.

Some years ago, I coordinated a national task force on sustainable communities for President Bill Clinton's Council on Sustainable Development, which brought together former adversaries—corporate and environmental leaders—to find common ground and build a consensus on environmental protection and economic development. This was probably the most interesting job I ever had, and its impact will continue through 2012.

The nonviolent change in the South African government and the transition from apartheid was facilitated by the groundbreaking work of the Truth and Reconciliation Commission—but also by the work of more than 10,000 people who were trained at the grassroots level in conflict resolution. Imagine the results if there were 10,000 trained facilitators in every war-torn nation.

Many of the breakthroughs in solving political conflicts around issues such as the environment or abortion are based on a common-ground, multistakeholder approach, where all parties with a stake in the outcome are invited to a professionally facilitated dialogue to find win-win-win solutions. A triple win means that both sides in a conflict, as well as the larger community, benefit from the outcome.

There is usually a grain of truth on each side of any political conflict. Healing, reconciliation, and forgiveness are spiritual qualities very much needed today. The impact of globalization is generating problems—economic, environmental, ethnic— that are too complex and interconnected to be solved on an adversarial basis. As Albert Einstein famously remarked, you can't solve a problem from the same level of consciousness that created the problem.

"Understand the differences; act on the commonalities," is the mission of Search for Common Ground (SFCG), the largest organization in this field, founded in Washington, D.C., by my friend John Marks more than 20 years ago. SFGC, like many similar organizations, offers a means of navigating through conflict and identifying possibilities that are not apparent from an adversarial mind-set. It draws upon the strengths of diversity and interconnectedness to find cooperative solutions.

All parties to a conflict are invited to the table and guided in how to shift from an adversarial stance toward a cooperative, problem-solving one. An essential step is enabling people to communicate and have accurate information about each other. Sharing stories and feelings helps those on each side of a conflict understand the pain suffered by each side.

Finding common ground is not the same as settling for the lowest common denominator or having two sides meet in the middle. It's about participants generating a new "highest common denominator" and identifying something together that they can aspire to and work toward, such as the health of children in a war-torn country. When those who really care about an issue come together and bring the best thinking from their various perspectives, there is the potential for new options to be generated—options that neither side might have thought of on its own.

It is especially important to help people involved in a conflict distinguish between positions and interests. Underlying someone's position on an issue are usually broader interests, such as security, respect, and/or the well-being of one's family. Interests can be discovered by continuing to ask "why" and inviting people to go deeper. Interests usually relate to basic needs, while positions are opinions about how to meet those needs. Positions may appear mutually exclusive, while interests tend to overlap, and this is the key to having all sides work together to transform conflict. SFCG calls this "cross-stitching," a way to reweave the whole.

It is also essential to distinguish between the problems and the people involved in a conflict. Helping people focus on common concerns, rather than seeing each other as the problem, is key. Rather than facing each other on opposite sides of the table, both parties are invited to sit on the same side of the table, and the problem they are addressing is placed on the other side. SFCG currently has more than 300 staff members working with local partners in more than 17 countries. For 10 years, they've sponsored high-level, unofficial (back-channel) meetings with Iranians and Americans, which have so far prevented war. SFCG and other organizations like it are the true unsung heroes without whom the world in 2012 would be in even worse shape.

Organizations such as SFCG use tools such as mediation, facilitation, training, and community organizing—as well as creative approaches such as sporting and music events with conflicting parties, and TV and radio soap operas—to teach ways to resolve conflict. They created the Common Ground News Service to feature solution-oriented articles to help resolve the Israeli-Palestinian conflict and bridge the gap between the West and the Muslim world. More than 2,300 of their articles have been reprinted in Arab as well as Western papers. Recently in the United States, they've been convening those who are on the extreme right and the extreme left to find common ground on the national health-care issue.

The Network for Life and Choice convened by SFCG and Public Conversations Project in Boston helped opponents on both sides of the abortion debate find common ground by redirecting their attention from arguing about when the fetus is life to instead looking upstream at conception. They discovered their common interests in wanting to prevent unwanted pregnancies and make adoptions more easily available.

Public Conversations Project has also been convening successful Arab-Israeli dialogues, environmentalist-developer dialogues, and Republican-Democrat dialogues. They ask, What if what unites us is more than we realize, and what divides us is less than we fear? They've recently written a guidebook called *Fostering Dialogue Across Divides*.

The Institute for Multi-Track Diplomacy, started in Washington, D.C., by Ambassador John MacDonald and Dr. Louise Diamond in 1992, offers training in conflict transformation techniques and helps resolve ethnic conflicts worldwide. The institute's approach is to involve all stakeholders—government officials, businesspeople, media, religious groups, nonprofit activists, and so on—in "multitrack" dialogues and to help people listen deeply to all perspectives. When people hear for the first time the personal stories and suffering of those on the other side, it often causes powerful breakthroughs in entrenched conflicts.

I served on the institute's board for more than seven years and was continually inspired by the amazing work it was doing in places as varied as Cyprus, Georgia, Kashmir, and Sudan. Through its work training several thousand Greek and Turkish Cypriots over the last 10 years, the gates between the two sides of the divided country were finally opened.

MULTISTAKEHOLDER DIALOGUES AND POLICY DECISIONS

Multistakeholder dialogues, which involve all parties in a collaborative conversation, are proving to be the most effective way to develop viable policies and reduce conflict on divisive issues such as development and environmental protection. These dialogues are convened by state and local governments (as well as by various nonprofit organizations, such as the National Coalition for Dialogue and Deliberation). By 2012, they could become a widespread approach to policymaking.

For example, a few years ago, I served as a small-group facilitator in an amazing experiment in electronic democracy organized by AmericaSpeaks. It was a "21st Century Town Meeting" in Washington, D.C., with more than 1,000 citizens building a consensus on an effective agenda for citywide policy for Mayor Anthony Williams. Each small group prioritized concerns and issues and fed them electronically onto a large screen that could be viewed instantaneously by everyone.

One by One, started by Suzanne Schecker, began facilitating dialogues between Holocaust survivors and former Nazis (and their children) and has moved into dialogues among adversaries in many other arenas.

Citizens in Chattanooga, Tennessee, transformed their city from the worst polluted place in America to a model of environmental sustainability through a series of community-wide multistakeholder dialogues facilitated by a professional team. All citizens were invited to a series of planning projects to envision what they wanted for the future. More than 5,000 ideas were generated and put into a computer and categorized. Everyone was invited back to a second meeting to see the patterns and relationships between the ideas and then to create a consensus on strategic goals. This culminated in a Vision Fair, celebrating the goals and inviting citizens to sign up for the goals that matched their priorities. The excitement of creating a communitywide consensus inspired everyone, including government, business leaders, and philanthropists, who helped make the visions a reality. Chattanooga attracted more than $800 million of investments in 223 projects, creating 1,500 new jobs and 7,000 temporary jobs.

TRANSPARTISAN POLITICS

In 2006, a study by the prestigious Princeton Survey Research Associates found that a remarkable 85 percent of Americans agree that the country "has become so polarized between Democrats and Republicans that Washington can't seem to make progress solving the nation's problems." The majority, 63 percent, wanted a bipartisan ticket with a prominent Republican and a prominent Democrat running together on the same ticket, collaborating.

The philosophical dividing line between the liberal and conservative positions is a disagreement over whether social problems are caused by economic factors or by a breakdown in individual values, and thus whether government or individual solutions are best. Conservatives argue that social problems are caused by personal choice and personal values, and so they see little benefit, for example, in government spending more on the poor.

Liberals argue that having good values doesn't help if there is not equal economic opportunity for all. Liberals accuse conservatives of coercive moralism, and conservatives accuse liberals of moral relativism that allows evil to flourish. The current policy deadlock in Washington is based on trying to separate economics and values from each other, with neither conservative nor liberal admitting any wisdom in the other's perspective.

One of the new political approaches that is gaining momentum is a synthesis of left and right. Synthesis is very different from compromise. Compromise is usually seen as the midway point between two polarities or two sides of an issue. Compromise is *not* the most effective way to deal with differences, as sometimes the deeper wisdom in each side is lost. Instead, the *best* in both liberal and conservative perspectives is highlighted on an issue to create a higher synthesis and more innovative policies.

Theorists such as Paul Ray, Ken Wilber, Drexel Sprecher, and Lawrence Chickering have been outlining some key components of a new politics beyond left and right. Today there are several emerging political groups working along these lines, and by 2012, these and other similar efforts may make a huge difference in political elections.

Reuniting America, directed by Mark Gerzon (who led the bipartisan Congressional dialogues) and Bill Ury (Harvard professor and author of *Getting to Yes*), aims to be what they call a "transpartisan campaign of political reconciliation." Their first Democracy in America gatherings included major leaders from both liberal and conservative national groups.

While strong on innovative processes for a new politics, Reuniting hasn't yet developed clear transpartisan stands on policy issues. But other organizations, such as Mark Satin's Radical Middle and Ted Halstead's New America Foundation, have been developing and publicizing innovative policy solutions that synthesize the best of left and right. The tagline for Radical Middle's online newsletter is "Thoughtful idealism; informed hope." Sociologist Dr. Paul Ray's research on Cultural Creatives and the New Political Compass shows that there's a big market for this new politics that is beyond right and left, and this could be well developed by 2012.

The sudden and surprising popularity of Illinois Senator Barack Obama and his book *The Audacity of Hope* may be one indication of this big market. Obama represents the next generation of leadership, which will be very powerful by 2012. He recently told *Newsweek*, "My peer group finds many . . . divisions [since the 1960s] unproductive. We see many of these problems differently—race, faith, the economy, foreign policy, and the role of the military. . . . The next generation is to some degree liberated from what I call the either/or arguments around these issues."

OUR CHOICE FOR THE FUTURE

All of these growing trends in business and politics are promising signs of a positive future that I believe could be more fully visible by 2012. They give me a real sense of hope for the future of democracy, as well as for our future economic and spiritual health, both here in the United States and around the world. Day by day, we are making choices in our own lives to either open to new ideas and embrace the values of this new world—such as compassion, connection, respect for differences —or stay stuck in fear, greed, or anger.

The choice is up to each of us. I'm choosing the new world, as I believe our lives depend on it. How about you?

The Alchemy of Time

UNDERSTANDING THE GREAT YEAR
AND THE CYCLES OF EXISTENCE

JAY WEIDNER

How do the ancient theories of cycles of time, like the Hindu Yuga system and four alchemical phases, figure into the 2012 prophecies? Jay Weidner has the answer. According to the Hindu Yuga system, we are in the Age of Iron, nearing the Golden Age. In the following essay, Weidner explores insights from the ancient art of alchemy that foretell how 2012 will shape our lives and our planet. Drawing from the codes embedded on the Great Cross at Hendaye, a 350-year-old alchemical monument, he reveals how the cycles of time, and alchemy, influence our fate.

> *The age of iron has no other seal than that of Death. Its hieroglyph is the skeleton, bearing the attributes of Saturn: the empty hourglass, symbol of time run out.*
>
> —Fulcanelli

The inspiration for this essay comes from my almost 19 years of research into the Great Cross at Hendaye and the French alchemist Fulcanelli. The unknown, elusive Fulcanelli, in his masterpiece *The Mysteries of the Cathedrals,* first brought the cross at Hendaye, France, to the world's attention.

While the details of this research can be found in other articles I have written and in the two books I coauthored,[1] it can be stated that the engravings on the Great Cross at Hendaye describe the end of not only the great four ages of the Hindu Yuga system but also the four ages of alchemical chronological timekeeping. According to the cross at Hendaye, the Iron Age, or the Kali Yuga, will be coming to an end with the galactic alignment on the winter solstice of December 21, 2012.

According to the mythology of the Yuga system, there are four ages of life and time on our planet. It is important to remember that in the Hindu Yuga system, there are many cycles within cycles, so the years I mention next may have to do with a larger cycle than the one addressed in this essay. Perhaps the huge number of years referred to below is the cycle of the solar system as it circles another star, or perhaps it is a portion of the great cycle of one revolution around the galactic center.

According to Fulcanelli, the first age is the Satya Yuga, or the Golden Age. The Hindu texts tell us that this age lasts 1,728,000 years. This is an age of extreme splendor, when the beings on our planet appear to have lived much longer than they do now. In this age, there are no wars, famines, strife, or evil.

The second age is called the Treta Yuga, or the Silver Age. As with the second law of thermodynamics, or entropy, in this age, things begin to slip, and the beings on Earth begin to deteriorate. This slippage, at least during this second age, is the beginning of corruption, and evil is introduced into the planetary sphere. According to the Vedic texts of India, this age lasts 1,296,000 years.

The third age of this cycle is called the Dvapara Age, or the Bronze Age. This is the beginning of the "fall" of humanity. In this age, corruption is more fully realized, evil begins to spread, and things start to fall into disharmony. This age lasts 864,000 years.

The last age is called the Kali Yuga, or the Iron Age. This is the age we are in right now. Evil and corruption become the driving forces, as greed, wars, famine, and disease spread across the planet like a tidal wave of death and destruction. According to the Hindu texts, this age lasts for 432,000 years.[2]

What is important to understand is that the Hindu texts are describing the four ages within a context of understanding that each successive age is shorter than the previous age. The Satya Yuga, or the Golden Age, is one-quarter longer than the Treta Yuga, or the Silver Age. The Treta Yuga is two-thirds longer than the Dvapara Yuga, or the Bronze Age. Understanding these differences in the lengths of the ages becomes important to the following discussion.

THE PRECESSION OF THE EQUINOXES

According to Fulcanelli and the cross at Hendaye, the alchemical four ages comprise the four quadrants of a 25,920-year cycle called the precession of the equinoxes. Essentially, the precession of the equinoxes can be explained by the fact that our Earth wobbles on its axis. Like a top spinning on the floor that loses momentum, the Earth's wobble takes nearly 26,000 years to unfold. The strange engraving of the four "A"s (see Figure 19) on the cross at Hendaye is a "hieroglyph of the universe," according to Fulcanelli. It is "composed of the conventional signs of heaven and earth, the spiritual and the temporal, the macrocosm and the microcosm, in which major emblems of the redemption (cross) and the world (circle) are found in association."

What Fulcanelli appears to be saying is that the precession of the equinoxes is to be divided into four distinct ages of 6,480 years each (4 divided into 25,920), which can be rounded to

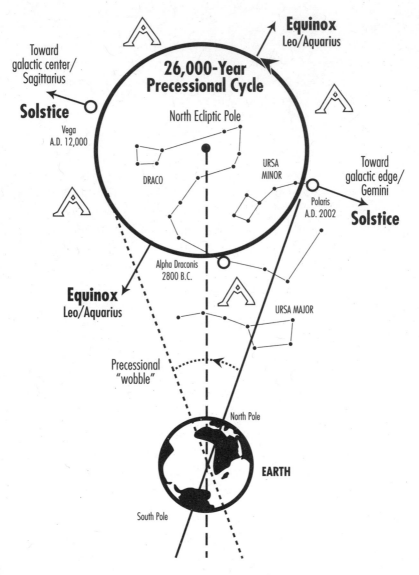

FIGURE 19: Precession of the equinoxes.

approximately 6,500 years. This is interesting because the zodiacal cycle, which lasts 25,920 years, has four fixed signs. These are the signs of Aquarius, Taurus, Leo, and Scorpio. These four signs are separated by 6,500 years. In the book of Ezekiel and in the book of Revelation, we are told of the angel with four faces. This angel has the face of a lion (Leo), a bull (Taurus), a man (the angel), and an eagle (in older times, Scorpio was represented by an eagle instead of a scorpion). According to Fulcanelli, both of these books are warnings and messages about the four quadrants of the precession of the equinoxes and are telling us that there will be a great change whenever we arrive at one of these four signs.

It is well known that we are currently entering the Age of Aquarius. The cross at Hendaye and Fulcanelli are telling us that there are tremendous changes to be found here on Earth when we enter into one of the four fixed signs. These changes can be cataclysmic.

Obviously, this 6,500-year dating of the four alchemical ages disagrees with the Hindu Yuga system, which insists that each Yuga is of varying time lengths. It is in trying to iron out the dissimilarities in these two time periods that the hyperdimensional aspects of time can be best revealed.

In modern physics, it is well known that four-dimensional space, of which time is an aspect, is in the shape of a hypersphere (see Figure 22 on page 207). A hypersphere is similar to the shape of a doughnut or a bagel. Physicists call this hypersphere a torus. The energy, or the flow of the energy stream, within a hypersphere works as follows: As shown in Figure 22, the flow comes from the bottom of the sphere, winds its way up around the outside edge, crosses the outside equator of the torus sphere, and moves toward the hole in the top of the torus. The energy flow then falls through the top of the hole and begins to spin in a

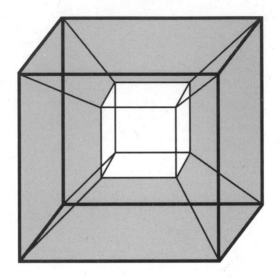

FIGURE 20: Hypercube.

vortex. This spiral energy flow continues until it comes out of the bottom side of the torus, where the energy flow renews its outward expansion. This flow is continuous and, in a sense, infinite.

The four-dimensional hypersphere torus is also represented by the hypercube. According to physicists, a hypercube is a cube within a cube. Essentially this cube, also called the "cube of space" or the "tetragrammaton" by the alchemists, is a straight-edged version of the doughnut-shaped hypersphere (see Figure 20).

Scientists love to construct objects using straight lines, so they have created the hypercube to help them better understand the hyperdimensional universe that surrounds us. Alchemists, however, like to use the curved lines of nature, so the doughnut-shaped torus or hypersphere is a better, more natural visual description of hyperdimensional space and will be the one I use in this essay.

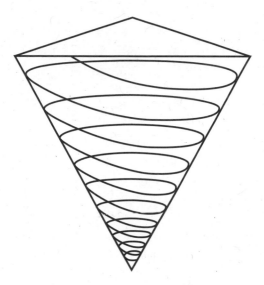

FIGURE 21: Inner tetrahedral vortex.

The alchemists believe there is a hyperdimensional "body" around every living thing on Earth. This hyperdimensional body or sphere, known to mystics as the luminous energy body, also surrounds every planet and star. Each object in the three-dimensional world is an affectation of four-dimensional space. Humans, animals, plants, planets, and stars are the solids inside the fourth-dimensional energy flow. As you will see, time also travels in this fashion.[3]

To comprehend the topographical nature of time, it is very important to understand the guts of this hyperdimensional sphere. By the "guts," I am referring to the vortexes that make up the energy flow running through the center of the hypersphere. As the energy flow begins to fall into the vortex that runs through the center of the sphere, it takes on the shape of a tetrahedron.

A tetrahedron (Figure 21) is the building block of three-dimensional space and is the founding member of the Platonic solids. The tetrahedron is the simplest object that can be created in three dimensions. The only object with just four sides, the tetrahedron is also shaped like a vortex.

Let us digress for a moment and discuss vortexes. In the Midwest, where I come from, there is a place called Tornado Alley, which extends across the Great Plains from Texas to the Dakotas. Each year, this area is visited by numerous tornados, which wreak havoc on the farms and towns of this area. The tornados usually start in May and last through early August. Having witnessed several tornados during my early years, I became fascinated by them and was once even a "storm chaser." In my young and foolish days, I used to chase after tornados with my Bolex 16 mm camera in order to capture them on film. I became very familiar with these lethal vortexes.

Understanding tornados also helps us understand vortexes and the density of forces in the hypersphere. Tornados (and hurricanes) are made up of the same air that surrounds us all the time. As the air begins spinning, however, it takes on a strange solidity. While the swirling air at the top of a tornado is somewhat dangerous, it is the spinning air near the ground's surface that is really dangerous. Near the tip of a tornado, where it is in contact with the ground, the air takes on the quality of solid iron. The tips of tornados can rip buildings apart, throw cars and trucks thousands of feet in the air, and punch chaffs of wheat like bullets into the very center of trees. As the air spins more violently, the tip of the tornado vortex becomes as strong as 50 locomotive engines and possesses the density of the hardest of metals. Yet, it is composed just of air molecules. These molecules are spinning so quickly that they create a mass that is extremely powerful. It is the spinning of the air that causes it to become dense and take on the solidity of metals.

Let's get back to the central core of our hyperdimensional torus. By using the tornado analogy, we can now understand the vortexes inside this sphere and how time unfolds. As the energy flow begins to dip into the top of the tetrahedral-shaped vortex, it begins to spin. As the energy flow descends, its spin becomes more rapid and compressed. Like a tornado, as it reaches further down, the tip of the vortex "hardens." It could be said that this hardening of vorticular forces is what makes up the solidity of our three-dimensional space. Each human, plant, animal, and indeed every planet and star alike are the hardened tips of hyperdimensional vortexes that are constantly flowing all around us.

Like air, these hyperdimensional forces are pretty much invisible to us. It is only when these forces coagulate into a spinning vortex that they can be seen and felt. It is through the rapid spinning of four-dimensional space that the solidity of the third dimension is actually created. The mystics from all great traditions know this, and they realized these inner dynamics of the fourth dimension as it interacts with the third dimension. The fourth dimension is the surrounding sphere, and the third dimension is made up of the tetrahedral vortexes in the central core of the sphere. The ancients called four-dimensional space "spirit" and three-dimensional space "matter."

What does this have to do with time and the topography of time? Remember the dilemma we faced in trying to make the four ages of the precession match up with the ages or Yugas from the Hindu system? By attempting to make the two into one, not only can we understand the topology of time, but we can also map time like the ancients did and then realize where we are in the river of time.

Observing Figure 22, imagine that we are coming through the very center of where the two tetrahedral tornado vortexes intersect in the inner hyperdimensional sphere. Moving down the bottom vortex toward the outside surface of the sphere, each spin in the vortex takes longer and is wider and slower than the previous spin in the vortex. In other words, the energy flow expands after it passes the "null point" in the very center of the hyperdimensional sphere, or torus, where the tips of the two tetrahedral vortices touch.

This energy flow continues its expansion as it climbs over the lip of the bottom edge of the sphere, where the vortex meets the outside edge of the sphere. The energy flow continues to expand until it reaches the outside equator area of the sphere. Once it gets past the equator, the energy flow begins to condense, the flow begins to move faster, and the spin density increases. This flow continues until the energy flow reaches the upper lip of the outside of the hyperdimensional sphere, where it begins to "fall" into the top of the tetrahedral tornado vortex.

Now, as the flow of energy falls down the vortex toward the null point at the very center of the sphere, it begins to harden as it spins faster and faster. This goes on and on until the vortex is as hard as iron. This is how the hidden fourth-dimensional forces create our visible three-dimensional reality. As the vortex spins faster and faster, it eventually compresses down to the point where it has nowhere to go but outward again. This occurs at the null point in the center of the sphere, and the energy flow begins to once again expand.

The Golden Age is the time period that starts at the null point in the center of the sphere and continues through the bottom vortex. As the energy flow expands outward, time appears to slow down. The entire period of the Satya Yuga, or the Golden Age, continues as the energy flow goes down the bottom vortex and rounds across the top of the sphere. It continues expanding until it arrives at the equator.

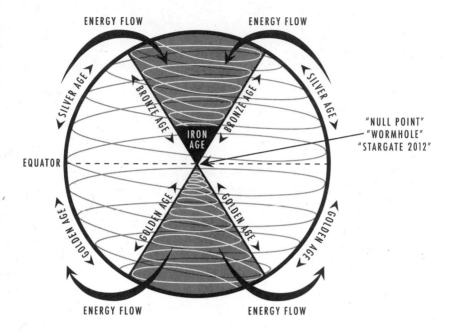

FIGURE 22: Hypersphere.

The equator of the outside of the hyperdimensional sphere is the borderline between the Satya Yuga and the Treta Yuga, or the Golden Age and the Silver Age. Here, the energy flow begins to contract as it flows up toward the north pole of the hypersphere.

The Silver Age, or the Treta Yuga, continues until the energy flow rounds the top lip of the hyperdimensional sphere and begins to fall into the upper tetrahedral vortex. This is the borderline between the Treta Yuga and the Dvapara Yuga, or the border between the Silver Age and the Bronze Age.

About two-thirds down the upper vortex, where the energy flow is spinning so fast that it becomes as solid as iron, is the beginning of the last age, the Kali Yuga, or the Iron Age. This spinning continues to gain density, compression, and speed as it races toward the central null point in the very center of the sphere. As it approaches this null point, the forces become unbelievably fast, violent, and dense. It is only when these forces achieve maximum density and they cannot compress any further that they begin to suddenly flip and begin the expansion of the flow. This happens in an instant. This is the shift from the Iron Age, or the Kali Yuga, to the Golden Age, or the Satya Yuga. The borderline between the Iron Age and the Golden Age is the most distinct border in this topographical illustration of hyperdimensional time. It is the most jarring, and it is instantaneous.

Take notice of the hourglass shape of the two tetrahedral vortices inside the hyperdimensional sphere. Is this why Fulcanelli tells us about the hourglass in the quote at the beginning of this essay? In Hindu mythology, the god Shiva dances the world into and out of existence playing his *dhamaru*, a two-sided drum shaped like an hourglass.

Notice that like the number of years in the Hindu Yuga system, the distance traveled through our ages in the hypersphere is similar; the Golden Age, which is the distance between the null point in the center and the outside equator, is twice the distance as that traveled through the Silver Age, which is the distance from the equator to the top of the lip of the upper vortex. Equally, the distance traveled from the top of the vortex to two-thirds down, which is the Bronze Age, is half the length of the Silver Age. Finally, the tip of the vortex is half the length of the Bronze Age.

The Iron Age is the age in which we live. This is why, in the Iron Age, each second feels shorter than the second before it. This is why each day, each month, and each year appear to be going faster than the ones that preceded them. In the topology of time, this effect can be easily understood and explained.

FIGURE 23: Great Cross at Hendaye and the four engravings on its base.

Remember, even though the distance traveled is much further in the Golden Age than it is in the Iron Age, the number of years it takes is the same. It takes 6,480 years to go through the Golden Age, just as it takes 6,480 years to go through the Iron Age, but it *feels* much different. In the Golden Age, each second, each day, each month, and each year appear to be longer than the previous day, month, and year. Time is expanding in the Golden Age, and with that expansion, the anxiety and tension of the Iron Age disappear. It is a paradise, especially for those who survive passing through the wormhole, the null point at the center of the hyperdimensional sphere.

Therefore, the years listed within the Yuga system are actually symbolic times that explain temporal passage as it is felt not as it is lived. So the Golden Age, or the Satya Yuga, feels much longer than the Kali Yuga, or the Iron Age.

The shamans in Peru tell us that we are approaching the Pacha Cuti, which means literally "the world turned upside down." They say that it will arrive in the year 2012. This is also the year when the great Mayan calendar reaches its end point. The cross at Hendaye is quite explicit in its depictions that we are reaching the very end of the Iron Age. I would suggest that the null point, or the wormhole at the very center of the hypersphere, is very likely the 2012 date. Our hypersphere appears to be regulated by the periodic alignment of the galactic center with the sun. This also matches the four fixed signs of the zodiac and explains why the Christian tradition is preparing for the end-times.

Oddly enough, it also appears that Stanley Kubrick was trying to tell us about the end of our age in his masterpiece film *2001: A Space Odyssey*. The star child at the end of the film has passed through the stargate of the black monolith and is returning to Earth to create a new world or a new age.[4]

It is through our understanding of the four ages, the fixed signs of the zodiac, and the Yuga system that we can finally appreciate the message our ancestors have handed down to us. We are about to pass through the stargate, the wormhole, and the null point of hyperdimensional space.

The Great Cross at Hendaye reveals the most important knowledge of all, which is the end of time and the beginning of the Golden Age. Awareness of this knowledge helps us all to comprehend more fully the nature of the times in which we reside and the destiny of the human race.

THE QUALITY OF TIME

At the very core of all the great spiritual traditions found across the planet—from the ancient Egyptians to the Incas and

Maya, Indo-Tibetans, alchemical Europeans, and the Hebrew Kabbalists—lies the knowledge of the secret of the alchemy of time. Unencumbered by the excessive demands, stresses, and speed of the modern age, our ancestors spent lifetimes contemplating the essential mysteries of our world.

One of the great mysteries they discovered was that time is not stagnant or linear but flows in great transformative cycles. In fact, it is only recently that historians have argued that time travels in a straight line. Within the limited period of historical examination, these modern historians are viewing a snapshot of history and believe this snapshot represents the entire picture. But does it?

The alchemists of Egypt and medieval Europe knew that time moved in great elongated cycles and that each part of the cycle had a quality. This quality shaped and reformed the world. Just as sunrise has a different quality from midnight, so, too, does each aspect of the Great Cycle possess a quality that distinguishes it from other parts of the cycle. When the cycle of time changes, so, too, do events on Earth.

The sacred scientists and alchemists of our past used various devices and practices to mark the changes in the quality of time. Astrology, divination, and prophecy are some of the more obvious processes that our ancestors used to understand time's mysterious process.

At the heart of their secret traditions, the Egyptian and Greek mystery schools knew that a process, which they called the precession of the equinoxes, measured the great cycle, or the great year. They understood that the stars in the night sky are not fixed. Through long observation and meditation upon the heavens, they came to realize that the stars and the constellations were slowly changing place in the sky. Creating the zodiac system to assist them in these huge measurements of time, they used the 12 signs as a giant clock. Through this they began to understand the great cycle. They realized that it took 2,160 years to move through each sign of the zodiac and that it took 25,920 years for the equinoxes to move through all 12 signs. The creation of the great year of nearly 26,000 years gave them a clock that allowed them to measure time.

They then came to understand that each of the signs, or places on the clock, appeared to bring forth changes in the quality of life here on Earth; they realized that every 2,160 years, the quality of time changed. They also began to understand that the four quadrants of this great zodiacal clock appeared to bring forth even sharper changes in the quality of time. The four signs associated with these quadrants are the signs of Aquarius, Taurus, Leo, and Scorpio.

There does seem to be a historical link to the idea that passing through the edges of these four signs brings great changes here on Earth. About 6,500 years ago, Earth passed through the sign of Taurus. Coincidentally, the human race began domesticating cattle, building walled cities, developing agriculture, fighting wars, and establishing the city-states that would evolve into governments and monarchies. Thirteen thousand years ago, as Earth passed through Leo, great changes occurred as well, including a dramatic climate shift as the ice ages came to an end. The extinction of many animals, including the woolly mammoth, the saber-toothed tiger, the giant tree sloth, and many other species, happened at about this time. Interestingly, according to anthropologists, the emergence of Cro-Magnon man appears to have occurred about 26,000 years ago—the last time we passed through Aquarius.

The Great Cross at Hendaye is a 350-year-old alchemical monument dedicated to the end of time. It sits in a churchyard in Hendaye, France. The Great Cross at Hendaye appears to be a symbolical construct that explains the science of the cycles of time. The cross describes the extraordinary changes that are about to occur in our present era. It speaks of the end of the current Iron Age and the creation of a coming Golden Age.

The great cosmic clock in the sky is moving. We are now entering the quadrant of Aquarius. The message on the alchemical cross at Hendaye indicates that this change is sure to have a dramatic effect on our species and on our planet.

What these changes are, or will be, is anybody's guess. They are sure to be spectacular and remarkable. The end of time, the end of this age, and the great transformation of humanity are at hand. This change in the quality of time will govern and dictate the changes on our fragile planet. The changes in the quality of time cannot be forestalled. The fundamentalist revival taking place all over the world reveals more about the quality of time from which we are emerging than it does anything about the quality of time that is coming.

The Peruvian shamans who live high up in the Andes mountains have come down to tell us that a new species of human being is about to appear. They call this being "Homo luminous." They tell us that forerunners of Homo luminous are already among us. Just like Cro-Magnon man appeared 26,000 years ago (the last time we passed through the cusp of Aquarius), so, too, a new human is now emerging.

This new human will possess entirely new qualities that we do not have. The great avatars—Jesus Christ, Krishna, and many others—may have been early arrivals of this newly emerging species.

Aquarius is the only one of the quadrant signs that is symbolically represented by a human, and this human carries a container of water. This symbology is interesting, as DNA can only be activated in water. Even the symbolic nature of the sign of Aquarius seems to suggest a change in the genetic structure of the human race.

It is interesting to note here that the human life span has a strange association with the hypersphere and the topology of time. It takes 72 years for the precession of the equinoxes to advance 1 degree (360 x 72 = 25,920). Interestingly, at least during the Iron Age, the average human life span is about 72 years, which is 26,280 days—very close to the precessional figure of 25,920. Incidentally, the ancients thought a year had 360 days, which would have made 72 years come out to exactly 25,920 days.

The human life span can also divided into four ages. If we divide 72 years into four periods of 18 years each, we can discover our own personal four ages of existence.

There are 6,510 days in each of these 18-year periods. A human's Golden Age is from birth to the age of 18 years of age. The Silver Age is the next period, from 18 to 36 years old. The Bronze Age of a life is from 36 to 54 years of age, and the Iron Age is from 54 to 72 years of age.

Like the four great ages of alchemical timekeeping, these four ages also feel different. It seems to take forever to turn 18 years old. But when we are 54 years old, or older, the days and years feel as if they are flying by. Also, the body appears to gain density as we age; our bones creak, and arthritis begins to attack. It becomes more difficult to lose weight, and we become set in our ways. Even our lives appear to be ruled by the topology of hyperdimensional space.

Our spiritual existence is a unique combination of free will and predestination. While we possess the ability to act freely within the moment, there is little we can do about the larger cycles of time and the qualities that emerge at the cusp of each of the four great signs. So, as the hand on the great cosmic clock passes into Aquarius, we can expect massive changes to occur. While the ancient art of alchemy is concerned with many aspects of our existence, including the extraction of light from plants and minerals, the transformation of the dark lead of our physical being into the pure gold of enlightenment, and much more, alchemy is also a symbolic ontology concerned with the observation of the changes in the quality of time through the 25,920-year duration of the great year.

The world is changing. The past will disappear, and history will become legend and finally myth. Our destination is unknown. The only thing we know for sure is that everything will change.

Mayan Stelae and Standing Stones

2012 IN OLD AND NEW WORLD PERSPECTIVES

JOHN LAMB LASH

The Mayan Long Count in relation to other ancient calendars is of interest to many 2012 debaters. John Lash, mythologist and author of Quest for the Zodiac, *explores 2012 as it relates to the modern zodiac and the Age of Aquarius. Lash questions such common assumptions through ancient myth and modern astrology: Are we actually entering the Age of Aquarius? How relevant is 2012 to the whole of humanity? Will Westerners be the ones who learn the greatest lesson?*

For the ancients, systems of sacred timekeeping were always connected to the stars. Stonehenge has been described as an astronomical clock constructed in stone. Its layout identifies the solstices and other factors of seasonal timing, including the 18.6-year cycle of lunar standstills—sophisticated knowledge, to say the least. An ascending passage in the Great Pyramid of Giza in Egypt was aligned to Thuban, the polestar, in 3100 B.C., around the start of the Mayan Long Count. Far from being a tomb, the Giza pyramid was used for rites of initiation, as were Mayan temples in the Yucatán. But there are striking differences between the sacred calendars of the old and the new worlds.

The Maya and Aztecs were accomplished star watchers who also designed pyramids as astronomical clocks—notably, the pyramid of Kukulcan at Chichen Itza, where light on its steps during the summer solstice produces a serpentine effect. By and large, however, the Maya recorded cosmic timing on stelae, or standing stones. In the old world, standing stones form layouts oriented both to the stars and to solar and lunar movements. Callanish in the Outer Hebrides, for instance, is aligned to the 18.6-year lunar cycle and possibly to Capella, the brightest star in the constellation of Auriga, the charioteer. Monte Alban in Mexico is also aligned to Capella, and other Mesoamerican temples show orientations similar to those in Europe. But in the new world, standing stones were reserved for recording dates, rather than for celestial alignments. The Mayan stelae do not form lines or circles oriented to the skies and seasons. The questions arise: What is the celestial factor in the Long Count recorded on these stones? In what way, if at all, does the calendric data engraved on the stelae incorporate astronomical timing based on the stars?

MIDNIGHT HOUR

In recent years, much work has gone into connecting the Mayan end-time with the 26,000-year cycle of precession. Before the 1970s, no one made this connection, because it requires locating the center of our galaxy relative to the zodiac. The galactic center is about 24,000 light-years from Earth and cannot be directly observed. The zodiac is the circle of 13 observable star patterns, a thin cross section of the sky (called the ecliptic) where the sun, the moon, and planets circulate. (This visible panorama should not be confused with the 12 astrological signs, which are uniform, starless sections of the ecliptic.)

The composite stars of the zodiac are packed into a sphere of visibility that comprises only 3 percent of all the stars in the galaxy. Nevertheless, looking out to that sphere, we can sight *through* the immense zodiacal constellations to the direction of the galactic center. In the 1970s, astronomers determined the sight line at 267 degrees on the ecliptic, close to the sting in the tail of the constellation Scorpio. Curiously, the neighboring constellation of the archer, Sagittarius, points directly to the cosmic locale by the tip of his arrow. Alignment of the winter solstice to this point *outside the zodiac* marks the midnight hour of zodiacal precession. With 3 degrees of precession to go, we are about 210 years from the midnight hour.

Precession in the zodiac is a measurable astronomical phenomena. The axial wobble of the Earth (like a top slowing down) causes the position of the sun at the spring equinox (March 21) to shift against the background stars at the rate of about 1 degree in 72 years. Due to this shift, over many centuries, the spring equinox, or vernal point (VP), slips into the preceding constellation, moving against the natural order of the zodiac; hence the term "precession." Throughout each year, the sun advances from Aries to Taurus to Gemini, and so forth. But over the course of centuries, the spring equinox shifts the other way, from Gemini to Taurus to Aries. Currently, the VP stands in Pisces, the constellation of the fishes, one of the largest in the zodiac. The VP is shifting slowly toward the urn of Aquarius, the Water Bearer. The parameters of this shift far exceed the end-date of the Mayan Long Count. (Figure 24: Alexander Jamieson, *A Celestial Atlas*, London, 1822)

Some scholars divide the full precession cycle into 12 intervals, wrongly called Platonic months. (In fact, Plato did not write on the zodiacal ages, and his "Great Year" is not the precessional cycle.) To assume 12 zodiacal ages of 2,160 years

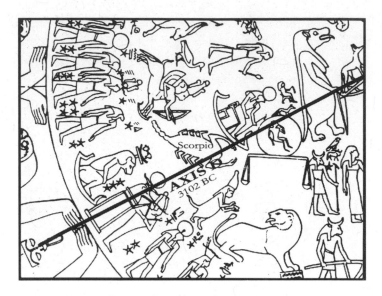

FIGURE 24: The vernal point in Pisces, 2007 C.E.

each (30 x 72) may be convenient, but this model ignores the real-sky environment observed by the ancients. In reality, there are 13 constellations in the zodiac, uneven in size and extent. To compute the ages by tracking how the VP actually transits these irregular patterns, rather than resorting to a contrived sequence of uniform intervals, is more consistent with how the ancients viewed the cosmos.

The Taurian Age extended from 4400 B.C. to 1850 B.C., the Arien Age from 1850 B.C. to 120 B.C. When the borders between the constellations are definite, transit dates can be computed with some precision. This is so for the shift from Aries into Pisces in 120 B.C., but not for the shift from Pisces into Aquarius. The transition of the VP into the composite stars of Aquarius has not yet begun and will take another 800 years. The dawn of the Age of Aquarius lies on the far horizon of human time. This shift does not match the Mayan end-time, but extends far beyond it. The preceding zodiacal ages, Taurian and Arien, do not tally with the Long Count, either, but there *is* one striking correlation between the Mayan calendar and zodiacal timing.

KALI YUGA

Precession functions like a millennial clock, so that every star in the zodiac can be precisely dated. The date 3100 B.C. falls in the Taurian age that began around 4400 B.C., when the VP shifted into the horns of the bull. In 3100 B.C., the VP, the hour hand of the zodiacal star clock, coincided with the golden star, Aldebaran, in the eye of the bull. Aldebaran is a *lucida*, a prominent marking star, one of five in the panorama of zodiacal constellations. It is located almost exactly 180 degrees across the zodiac from another lucida, Antares, in the heart of the scorpion. The opposition of Antares and Aldebaran defines the structural axis of the entire zodiac, as attested by diverse ancient astronomical myths and megalithic alignments. Aldebaran sets exactly when Antares rises, and vice versa. Sidereal lore, such as the myth of Orion the Hunter, encodes the movements of these two uniquely counterpoised stars.

The Hindu Yugas are linked to the Long Count by a coincidence noted by many scholars: the initial moment of the Long Count in 3114 B.C. falls close to the date 3102 B.C. recorded in Hindu mythology to mark the death of Krishna, the divine avatar who delivered the wisdom discourse in the *Bhagavad Gita*. In Hindu chronology, 3102 B.C. marks the start of Kali Yuga, the age of ignorance and degeneration in which we are now living. Kali Yuga tallies with the Long Count *at the start date*, but Hindu systems give the Yuga a longer duration than the 5,140 years of Mayan reckoning. For what it's worth, this is the Hindu-Mayan correlation.

Another clue to the astronomical timing of the Long Count comes from Egypt. The Dendera Zodiac, from a temple on the Nile north of Luxor, presents the oldest, intact working model of the zodiac that survives from antiquity. Its five axes record dates in prehistory and history, including 3100 B.C. At its southern end, Axis D at Dendera intersects Antares in the heart of the scorpion. Egyptian myth recounts the murder of Osiris, a parallel to the death of Krishna in Hindu tradition. The Egyptians pictured Osiris, called Asar in his celestial form, in the constellation of Orion the Hunter, a figure closely associated with the Antares-Aldebaran axis, as already noted.

The Hindu-Mayan-Egyptian correlation locks the Mayan Long Count into the frame of cosmic timing measured in the zodiac, but it doesn't yield a decisive end-date in astronomical terms. The "lesser" Kali Yuga does not end in 2012, but in 2082, and the Dendera Zodiac encodes the terminal moment in 2216, when the winter solstice aligns to the galactic center—the midnight hour of the full precessional cycle of 25,920 years. With a discrepancy of more than 200 years, the Long Count cannot convincingly be correlated with the midnight hour of the zodiacal ages. Placed in the frame of the ages, the Long Count fits in some ways, but the Mayan end-time just does not compute consistently in astronomical terms. (Figure 25: Dendera Zodiac Axis D, after Schwaller de Lubicz, *Sacred Science*.)

NEW WORLD KARMA

There are, nonetheless, valuable lessons to be drawn from considering the Long Count in the zodiacal perspective. The Mayan end-time will not be the close of Kali Yuga, the age when humanity degenerates into stupidity, depravity, and violence, and it will not be the dawn of the Aquarian Age, which still looms on the far horizon of human imagination, but it does

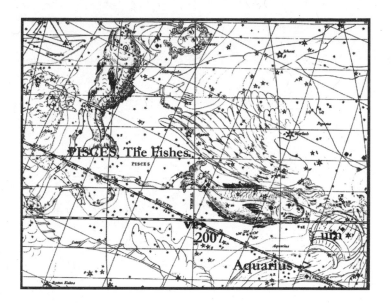

FIGURE 25: Axis D in the Dendera Zodiac.

represent a unique moment of reckoning in the overall pattern of zodiacal timing, perhaps signaling a critical turn for Western civilization. The standing stones of Europe are unmarked, rugged sentinels oriented to cosmic timing, erected in prehistory. The stelae of Central America are ornately carved pillars loaded with calendric, tribal, and historical data, and they were erected in historical time (at Izapa around 250 B.C., according to John Major Jenkins). These are essentially different media encoded with quite different messages. The Mesoamerican stelae carry a heavy burden of historical karma.

As a cultural artifact of the Americas, the Mayan Long Count differs from megalithic timing systems in the old world. In *Mexico Mystique*, one of the earliest books on 2012, anthropologist Frank Waters remarked that the Mayan-Aztec calendar system has a unique relevance for the Americas where it was created. The 2012 end-time may be a wake-up call for the entire planet, but it primarily implicates those who are involved in the excessive Western way of life exemplified by the United States.

America leads the world in pollution and consumption of natural resources, not to mention dirty dealings in imperialist aggression and profiteering (thinly disguised as global police work and the promotion of democracy) that bring death and anguish to untold numbers around the world. Mayan-Aztec calendrics cannot be dissociated from two alarming dramas in the past: the disappearance of the classic Maya after the tenth century and the eruption of the Aztec apocalypse in the thirteenth century. The rapid disappearance of an entire civilization, or its self-destruction through obsession with retributive and sacrificial violence, belong to Western history and may be repeated. The psychohistorical legacy of ancient Mexico, to which the peoples of the Americas are the main heirs, could determine the fate of human life on the planet, for better or worse. The 2012 end-time may be *the* point of no return for the Western way of excess and exploitation. It may also be the moment of opportunity for some people in the West to set a course correction for the entire planet.

Like the Giza pyramid and Stonehenge, the Mayan Long Count comes without a set of operating instructions. Whatever happens after 2012 will be largely, if not wholly, determined by the corrective and creative intentions brought to that moment. Apart from the overruling powers of nature, nothing decides the outcome of the end-time except the will and imagination of those forward-seeing people who can understand and actualize this rare mythic opportunity.

PART THREE

Spirituality, Signs, and
Symbolism Surrounding 2012

The Clock Is Ticking

ARJUNA ARDAGH

*Some believe that 2012 heralds a time when the majority
of humanity will be awakened to a new consciousness or unified
spirituality. In some respects, it seems like a great leap
from where we are now, and immediately one might wonder,
How is that possible? Arjuna Ardagh, author of* The Translucent
Revolution, *among other books, has done a great deal of work
investigating modern personal awakenings, interviewing hundreds
of people about their experiences of breaking out of the "trance
of separation." In the following essay, he shares his findings. How
can everyone possibly find awakened spirituality and function as
a unified harmonious humanity by 2012? Through his work with
the Oneness University, a possible catalyst for both personal
and collective awakening, Ardagh may have found a way.*

Not so long ago, I received a pamphlet in my mailbox. "In a
world gone slightly mad . . ." read the cover. It was asking me
to subscribe to a new alternative magazine. It was a very clever
marketing hook, because most people would feel immediately
that it was "talking to them." We may have different political,
environmental, or economic leanings, but today, most everyone
agrees that we live in a world gone slightly mad. Some see the
core of the problem as sinister terrorist plots and feel we can

only relax when we have killed the very last "enemy combatant." Some see the root of the decay as rampant and blind imperialism and seek a change of government. Some see the madness as a decay of moral values and believe the solution is a return to family values in the media. Some see the writing on the wall as an impending environmental crisis: we're busy pumping black sticky stuff out of the Earth as fast as we can; our whole economy runs on it; it has to run out sooner or later—and then what? Other people feel, quite intelligently, that burning all this black sticky stuff, whether it runs out or not, is heating up the planet so fast that we are making ourselves extinct. Those are just a few of the possible ways to see that the world has gone slightly mad.

If all this were not enough, a report published in the winter of 2006 tells us that 50 percent of the money in the world is now owned by 2 percent of the population, and 40 percent is owned by 1 percent. Imagine this is like Thanksgiving dinner. Your whole family is there: parents, children, uncles, aunts, cousins, in-laws, your closest friends. Look at your plate. It is piled high with turkey, potatoes, stuffing, ornately carved vegetables, and rich gravy. You have several crystal glasses filled with different European wines. You have an exotic seaweed salad, imported from Japan. On a nearby table, there are many different kinds of dessert—all for *you*. The people seated closest to you didn't do quite as well. They each have a microwave turkey dinner—$3.99 from the grocery store. And Coke. The rest of those at the table really missed out. Some have bread or gruel and water—contaminated. Some have nothing. Could you enjoy your meal under these conditions? Surely not; you would have to share. Having Thanksgiving dinner with your family when one person has everything and other people have nothing would be insane. And yet, we concede that this is an acceptable way to live as a global family.

On the surface of things, it looks like all our problems are disconnected from each other. We've got to deal with terrorism, find a way to elect saner leaders, stop global warming, find alternative energies, clean up air pollution, and stop the juggernaut of consumerism, if we want to survive. Each of the challenges we face demands its own solution. The only thing they all have in common is that we don't have the first clue what to do.

Many years ago, Albert Einstein said, "You cannot solve any problem in the same state of consciousness in which it was created." So, on the one hand, all of the challenges we face on this planet seem separate; on the other, they are all symptoms of a state of collective consciousness that is no longer viable.

What is that state of consciousness we have come to regard as normal, even if the heart will not accept it as *natural*? Pigs grunt. Dogs bark. Penguins waddle. What do human beings do? What have we come to accept as the default state of being human? I've spent many years researching that question, asking writers, teachers, psychologists, even hairdressers. Most agree that human consciousness is characterized by an unnatural sense of separation, a sense of a "me" and a "not me." We act as though the source of thought within us and the source of everything else are separate; ultimately, we act as though we are separate from the source itself, from the divine. On the basis of this feeling of separation stands everything else that feels abhorrent to the heart—child abuse, domestic violence, people lying to and cheating each other, environmental degradation, war. All of these things arise from this feeling of "me" and "them" as separate, or "me" and "the planet" as separate.

Because we feel cut off from all of life, our relationship to the world is characterized by desire. We often feel that something is missing; if only we could get what is missing from the outside, then we could relax. This is the very mind-set that propels the global economy. This is how it is possible to sell more and more stuff to more and more people. "You think you're missing something, buddy? Yep, you sure are. Now, if you just buy this new barbecue here, then you'll feel happy. Just look at these people in this here picture; they're happy. You could be happy, too, if you buy this barbecue. . . ." Or this car . . . or this vacation . . . or this new house. When people feel lack, they hoard things, and then there is no flow.

While this sense of lack may sometimes be caused by real hardship, most human beings live in "psychological lack:" a feeling that they need more money, more things, more power over more people, more sex with more partners in a greater variety of positions. They just need *more.* Then they can relax. At one time, Paul Getty was the richest man in the world. Toward the end of his life, he was asked in an interview, "Mr. Getty, you have more money than anyone in the world. When is enough enough?"

"Not quite yet," he replied. It's always like that—not quite enough, not quite yet. As human beings, we have learned to run on this "sense of lack." It causes us to live in endless worry, a disease of excessive thinking.

In every culture and in every age, a few isolated individuals have broken free of this hallucination and have realized that the sense of a "separate me" is actually a fantasy. This is not a question of self-improvement, or working on yourself to make yourself into a more loving, conscious, better person. It is a sudden and radical shift from a preoccupation with "me" and "my story" to a realization of the space, the vastness, the eternity in which that story is occurring. Once this core hallucination is corrected, everything else starts to fall into place. Relationships, for example, are no longer between a "me" and a "not me." Now it is possible to look into the appearance of the "other" and to recognize that the presence being met is the same as the one who is looking, albeit in a different package. The sense of a separate entity that craves and needs things from its environment starts to melt like a snowflake in the sunshine. In its place is recognized consciousness, presence, a causeless love that lacks nothing at all. Rather than being on this planet "to get more for me," life becomes a flow of generosity of spirit. The reason to be alive is to bless, to love without restraint, to give irrationally.

There have always been people, here and there, who have passed through this shift of identification from form to formlessness, from content to context. Two and a half millennia ago, for example, there lived a prince named Siddhartha. For many years, he had tried everything available in the sphere of working on oneself. Finally, he got fed up. He sat down under the nearest tree, which just happened to be a Bodhi tree. During the night, all kinds of hallucinations visited him in his mind: bizarre sexual fantasies, imaginings of great wealth and power, images of great fear. But he had the sanity and the presence to know that they were all the products of his mind. As the sun came up in the morning, he came to a simple realization. He saw that all of this—every thought, every memory, every desire, and every fear—is just the play and activity of the mind; it is all fundamentally unreliable. He asked himself what remains when one is free of clinging to mind forms. He realized that his true nature was awareness, consciousness, both formless and empty. In English, we say "awareness," but the word for this in his language was *budh*. And so it was that from the day of that simple recognition he became known as one who is that budh, or Buddha. It is reported that when he returned to his friends in Sarnath, who had been his companions in his seeking, they recognized a radiant glow in his face. "Whoa, what happened to you?" they asked. "You look great! Did you find a new practice or a new teacher? What did you do?"

"I did not *do* anything," he is said to have replied. That is the beauty of it. Nothing changed. Nothing was improved. There was simply the recognition of that which is deeper than all change.

In the last two decades, there has been an explosion, all over the world, of people having direct realizations very similar to that which Buddha recognized under the Bodhi tree—that what I truly am is not only Bill, or Cynthia, or Robert, but what I really am is budh: awareness, consciousness, presence. This realization may come as a snapshot out of time and then be overshadowed by the pressure to pay the rent. It may come as a more sustained opening. It may even become the very fabric of day-to-day life. But ultimately, it does not really matter. Once the truth has been seen, the game is up on the hallucination of separation. The undying allegiance to the seductive stories told in the mind has been broken, and something more sane, more present, and more stable has a chance to shine through the habits of personality.

I call people who have been transformed in this way "translucent," which Webster's dictionary defines as "letting light pass through, but not transparent." A transparent object, such as a clean sheet of glass, is almost invisible. You see everything through a transparent object as if the object were not there at all. An opaque object such as a solid wall, on the other hand, blocks light completely. A translucent object allows light to pass through, but diffusely, while maintaining its form and texture. Objects on the other side cannot be clearly distinguished. A crystal is translucent. So is a sculpture of frosted glass: if the sun were to shine on it from behind, you would see the light passing through the sculpture, and the sculpture would appear to be glowing from the inside.

Translucent people also appear to glow from the inside. They have access to their deepest nature as peaceful, limitless, free, unchanging, and at the same time, they remain fully involved in the events of their personal lives. Thoughts, fears, and desires still come and go; life is still characterized by temporary trials, misfortunes, and stress. But the personal story is no longer opaque: it is now capable of reflecting something deeper, more luminous and abiding that can shine through it.

Over many years, I have conducted hundreds of in-depth interviews with contemporary translucents and have surveyed 13,000 more. Those interviews fill 250 cassettes; the transcripts make up more than a million words. They include such well-known writers and teachers as Eckhart Tolle, Byron Katie, Ram Dass, and Jean Houston. But I also talked to dentists, hairdressers, housewives, and hobos. I've interviewed politicians, drug dealers, and my tax consultant. I regularly survey whoever is sitting next to me on an airplane. Across the world, from every imaginable background and system of belief, people report that the trance of separation is being broken. And, as sociologist Paul Ray shared with me, "the mouth of the funnel is getting bigger every day." According to the most conservative estimates, millions of people today have been touched by such an awakening, most of them within the last 15 years. My 2005 book, *The Translucent Revolution*, is the fruit of that research and presents a portrait of what may potentially be a new kind of human being.

The world today is not the same world that Buddha lived in; it is not the same Earth that Jesus walked on. Life has changed. While the realization may be essentially the same, the challenge of embodiment is now quite different. How does this kind of awakening transform ordinary life? What is our relationship to thoughts and to action when we know ourselves to be the consciousness in which those thoughts are arising? What is our relationship to feelings, like anger or jealousy, when we realize that these feelings are passing through us and are not who we are? How is it to be in relationship? How is it to be sexual? How is it to be a parent? How is it to be an artist, to run a business and make money? How is it to be a priest or a bishop or a rabbi in the light of this awakening?

When I set out on this somewhat overambitious research project, I harbored the mistaken idea that this epidemic of awakening was limited to recovering New Age hippies like me. People with a leaning toward India, brown rice, and ponytails. I was inspired and relived to find out that no one has a monopoly on this shift in consciousness.

I interviewed the retired bishop of Edinburgh Cathedral, Richard Holloway. We are not talking Findhorn here: Edinburgh Cathedral is a stronghold of protestant conservative values. Bishop Holloway told me, "It's such a relief to find out that the prison door was never locked. That the Jesus I was praying to can be found behind my own eyes and in the eyes of everyone I see." Those are not the words of Ram Dass or Eckhart Tolle; that was the Bishop of Edinburgh Cathedral. I've interviewed people who work in inner-city schools in New York City, consultants who've worked with the Dutch company Unilever, the largest manufacturer of consumer goods in the world. I was shocked to find out that the senior managers of Unilever are all "falling awake." They're going on retreats together. They've changed so many things about their company in light of this shift of consciousness. I found environmentalists, doctors, politicians, artists, and housewives who have all been touched by the eternal in a way that has transformed the way they live. By the end of my long period of research, no doubt remained in me that however severe our challenges as a race, we are also in the early stages of a huge shift in collective consciousness, which is gaining momentum every day.

Our situation on this planet today could be compared to a James Bond movie: The plot always follows the same basic template. A "bad guy," usually very rich, greedy, and slightly mad, has malicious intentions to take over the world, causing widespread destruction in the process. As I recall, all the "bad guys" were men, with names like Dr. No or Goldfinger. He usually heads up some sort of global clandestine corporation, often with a benevolent facade. His operatives are dark, inhuman, and robotic. Then, there is Bond—suave, centered, in his body, living totally in the moment, and free of fear. He is humorous, even in the face of death, and brings a lust for life to every situation. Again and again, against all odds, Bond saves the world. It's usually a race against time: Will Dr. Destruction detonate his mother-of-all-bombs and turn us all into plasma, or will James single-handedly wrestle hundreds of meanies to the ground and save the day? Bond always comes through, with seconds to spare, and the movie ends as Bond has yet another liaison with a goddess-of-the-day.

In one way, these far-fetched scenarios have prepared us for the situation we face at the start of this new millennium. The threat to our world's stability comes from a collective expression of greed, most pointedly embodied by global corporations, which put profit before integrity. Like Bond, translucents are heavily outnumbered. But also like Bond, they are sleek, sexy, cool-headed, humorous, and increasingly motivated toward social change. So, here we sit, on the edge of our seats, at the last and most gripping act of the movie. Will we see life as we know it irreparably mutilated by corporate greed and fundamentalists bent on proving themselves right and the enemy-of-the-month wrong? Or are we finally at the dawn of a collective shift to sanity? No point in twiddling our thumbs in anticipation; the final pages of the script are still being written, and you and I have been handed the job of finishing it off.

So, what is the timescale for this global thriller to come to its conclusion? How much time do we have? I have put this question to a great many world-renowned authorities. Several environmental reports on resource depletion, the disturbance of weather patterns due to global warming, and the melting of the ice caps have targeted the start of the second decade of this century as the time when it will be impossible to overlook the damage already incurred.

Peter Russell has extrapolated that with the increasing speed of computer connectivity over the Internet and the increasing number of people with access to it worldwide, the world is becoming interconnected much faster than anyone anticipated. Within the first years of the next decade, Russell feels we can postulate the emergence of a "global brain." Every one of the 10 billion cells in the brain has the capacity to connect to every other cell in almost no time. When we combine the increase of Internet connectivity, the speed of connection, and the increase of computer speeds, we can see the real possibility of this global brain becoming reality in the next years. Russell describes this planetary consciousness as "billions of messages moving backward and forward in an ever-growing web of communication, linking the billions of minds of humanity into one single system."

Concurrently, the number of people affected by a radical awakening, as well as the depth of that awakening, is growing every day. In a sense, the worse the outer situation becomes, the more incentive people have to wake up from the dream of separation and desire. It is difficult to count, with any accuracy, the number of translucents worldwide. They do not register themselves or conform; on the contrary, they tend to walk independent of any sense of membership, more like lions than sheep. Sociologists and visionaries like Paul Ray, Barbara Marx Hubbard, and Duane Elgin estimate that the number of people whose lives have been touched deeply enough by an awakening to have had a lasting transformative effect to be in the millions, and this number is increasing exponentially. By the early part of the next decade, it could reach 60 to 70 million, or 1 percent of the world's population.

Perhaps by no coincidence, predictions in many traditions also point to the early years of the next decade as a time of a quantum shift. Some emphasize the collapse of many human-made systems we have grown used to; some emphasize the emergence of a new, more evolved human being, free of the crippling sense of separation we think of as normal. The Mayan calendar, for example, sees the end of psychological time as the year 2012, after which we enter into a new age, free of mind.

In *The Romeo Error*, Lyle Watson was the first to suggest that it might take only 1 percent of a population to undergo a shift toward greater coherence for the whole population to shift. More recently, Malcolm Gladwell wrote *The Tipping Point*, in which he credibly demonstrates how shifts in the collective start at the local level. He shows how one man who took a stand against gang violence in New York City was pivotal in a dramatic drop in the crime rate of the whole city; how just a few people can start a new fashion for shoes that takes over the country. It only takes a few people, the right people, to change isolated events into an epidemic. Research in marketing trends, for example in the shift from cassettes to CDs as the preferred media for music, again and again shows a quantum step of gradual growth to a "tipping point," followed by a sudden jump to universal acceptance.

When Nicolaus Copernicus suggested that the Earth was a planet among others, all of which were orbiting the sun, his theories were virtually ignored in his lifetime, and his major work was not even published until the year of his death in 1543. It took Tycho Brahe, Johannes Kepler, Galilei Galileo, and later Sir Isaac Newton to prove that his theory was correct. They did not have to travel door to door, pitching their realization to all of renaissance Europe. It only required a very few intelligent people to test their theories, and within a few years, one paradigm was completely replaced by another. Despite great resistance from those in power, the theories of these astronomers eventually prevailed because they were true and because they were verifiable through open-minded inquiry.

We are now at the beginning of another paradigm shift, one just as powerful. Copernicus's theory was about the physical universe and laid the foundation for modern science. The epidemic of awakening that is just gaining momentum is about human consciousness and could lay the foundation for an evolutionary leap in human life unlike anything we have known.

I finished *The Translucent Revolution* with three possible scenarios, each of which could easily transpire in these next critical years leading up to 2012. The first, which some people take quite seriously, is the possibility of our complete extinction. In this scenario, the fundamentalists over here and the fundamentalists over there finally lose all restraint. They all drop their bombs, and that's the end of human beings, and everything else. We can't rule this possibility out completely, although it seems unlikely. Life usually finds its own balance.

I call the second scenario "extreme crisis." The shift in weather patterns due to global warming, our running out of oil, and the extreme economic disparity between people could cause a collapse of organized society, whereby humanity survives but has to pass through a period of breakdown.

I call the third possibility "the miracle." Somehow, through some kind of divine intervention, everybody wakes up really, *really* quickly. Almost overnight, we all shift from burning oil to using alternative energy. We cease to indulge wars of fundamentalism—the energy and creativity liberated in the process is enough to repair the ozone layer and take care of global warming, and everyone takes a deep breath and relaxes. This might seem ludicrously optimistic to most people. It seems it would require the intervention of something beyond human consciousness.

One might wonder about the crop circles, for example, which have appeared in western England, among other places. Who made them? As their complexity has increased, the possibility of a human hand being responsible for them has diminished. Whatever made them seems to be highly intelligent, have a sense of innate beauty, and also be very powerful. Might whatever created art in a cornfield also help us to wake up? When I finished my book in 2004, I was, like many other people, waiting and hoping for a miracle.

Soon after the book came out, my wife and I received an invitation from the Oneness University, an organization in the south of India. The founder had read my book and invited us to visit. Initially, we were resistant, both of us having already spent a good part of our lives influenced by Indian teachers and teachings; but one thing led to another, and then another, and we wound up accepting their invitation.

What we discovered there may well be the kind of miracle that many of us have been waiting for. Over the past few years, the phenomenon that started at the Oneness University has spread all over the world, grabbing the attention of some of the most intelligent authorities in collective consciousness. In India, it is known as "deeksha," but in the West, it has come to be known simply as the "Oneness Blessing" and has proven to be, in the opinion of many respected writers and teachers, the most effective catalyst for both personal and collective awakening that they have ever seen. The Oneness Blessing is a simple transmission, most commonly given by blessing-givers placing their hands on the recipient's head for fewer than two minutes.

The particular clue that the Oneness Blessing offers to our collective situation is the degree to which awakening is a function of neurology. It is possible to have all kinds of spiritual insights, to do all kinds of work on yourself, but if the balance of the neurological activity in the brain is not able to support those insights, they don't stick. The Oneness Blessing brings changes in the balance of activity in the brain, which allows it to support awakening in a spontaneous and effortless way.

The Oneness Blessing has only emerged quite recently, so the investigation that has been conducted at the time of writing is still in its preliminary stages. Early speculation based on readings using EEG and SPECT scans, as well as more unconventional measurements using devices like Karnak and Amsat, all point to the same five preliminary conclusions about how the Oneness Blessing affects the brain. First, it is claimed that there is reduced activity of the brain stem, sometimes called the reptilian brain, widely accepted to be overactive in contemporary humans. Bent on survival, it is an inappropriate part of the brain to be making decisions about how we live. Second, preliminary measurements have shown a balancing of hormonal secretions governed by the limbic cortex. Third, many investigators claim that there is a significant reduction of activity in the parietal lobes, sometimes called the Orientation

Association Area. Fourth, it is claimed that there is an increase in activity in the frontal lobes, particularly on the left side. And last, there is an increase in whole brain coherence, indicated by increased neurological activity in the corpus callosum.

Dr. Andrew Newberg is considered one of the world's leading experts on the brain and higher states of consciousness. Through research on the effect of meditation practices on long-term Tibetan meditators, his team at the University of Pennsylvania has documented many of the same shifts discovered with the Oneness Blessing. He considers the parietal lobes to be very overactive in contemporary human beings—possibly as much as three times. While their healthy function allows us to negotiate three-dimensional space, when they are as overactive as they are in most human beings, it creates an exaggerated and hallucinatory sense of separation from one's environment. When the parietal lobe activity reduces beyond a certain threshold, the sense of a separate self dissolves, and there is the natural realization of "oneness," which Newberg calls the experience of Absolute Unitary Being. His research also shows an increase in frontal lobe activity, particularly on the left side, which he feels is associated with greater presence and a sense of well-being, or bliss, with no outer cause.

What makes the Oneness Blessing remarkable is that these same changes in neurological functioning, as well as correlated changes in the subjective experience of reality, happen in a matter of minutes rather than over several decades. Furthermore, the effects are cumulative. Pilot research studies show that with even a single Oneness Blessing, parietal lobe activity reduces significantly, and frontal lobe activity increases. After a week, these effects are not completely lost, and with further Oneness Blessings, the changes in the brain continue to deepen and stabilize. With enough repeated administrations of Oneness

Blessings, it is suggested that the brain returns to a state of healthful balance and then remains there.

Millions of people have received the Oneness Blessing since it started to spread globally in 2004. Thousands of people appear to have already shifted into a stable shift in brain functioning, and its growth continues exponentially. All of this is not happening randomly. The Oneness University was founded by a couple, originally schoolteachers, now known as Sri Amma and Sri Bhagavan. They have had visions for the potential of this phenomenon, shared by thousands of people from all over the world who have trained to give the Oneness Blessing. They see that as the Oneness Blessing spreads, it will precipitate a tipping point: enough people who shift into an "Awakening into Oneness" will be the catalyst for a shift in the collective. They see all this as happening not in decades but in just a few years. The date they see when the collective consciousness of humanity will have shifted from separation to oneness? The year 2012.

Wild Love Sets Us Free

GILL EDWARDS

As discussed in Arjuna Ardagh's essay, many believe that the spiritual change that will happen around 2012 is one of hope, oneness, and unity. This phenomenon is often called a "great awakening." Gill Edwards, a clinical psychologist, spiritual teacher, and best-selling author, says the key to arriving at this great awakening is very simple—we must learn to embrace what Edwards terms "wild love." She shares her view on the forecasts of chaos and disintegration leading up to 2012 and answers the question many are asking: What can we do, in our everyday lives, to prepare for the coming change?

Last night, I dreamt that I was a deer—roaming free through the forest, leaping over bracken and streams, pausing to graze in a woodland clearing, my ears twitching and alert, my eyes and nose alive to everything around me. I awoke from my dream, rose from my bed, and opened the shutters. Morning light poured into the room, and I gazed into the garden. A few paces away, beneath an oak tree, a deer looked up and glanced toward me. For a split moment, I felt my ears twitch and did not know whether I was deer or human.

In the Native American tradition, the deer is a symbol of gentleness and unconditional love. It reminds us that love is the great healer, which overcomes the illusion of separation. When love is present, there is no fear. When there is no fear, everything is welcomed into your heart—and you relax deeply. Love reassures us that all is well. It sets us free. Love unites and connects, whereas fear disconnects. Darkness is merely the absence of light—and, like the morning light, love pours in unless we are shutting it out.

A NEW COSMOLOGY

For me, this time of great awakening (that is, the decades leading up to and beyond 2012) is all about love. It is about expanding our consciousness beyond the old creation myths and realizing that the cosmos is based upon unconditional love, from which we are inseparable. Love is what we are. What is more, we are not victims of fate or chance or karma, but divine cocreators of everything (without exception) that happens to us. We are creative fragments or sparks of Source/God/Love. This loving universe is primed to help us make our dreams come true, and we are apprentice gods and goddesses, who can allow our dreams in or not.

At the moment, many of us are straddling two different worlds: the commonsense world of fear, struggle, blame, and victimhood that is mirrored by the news media, and the emerging world of love, joy, healing, and spiritual awareness that is bubbling up behind the scenes, within the ever-expanding group of "cultural creatives" who are holding a new consciousness.[1] Once we step fully into this new reality, everyday life takes on a numinous quality, and we feel more and more alive. We realize that the outer world is a mirror of the inner world. There is no real separation between inner and outer—just as the deer in my dream then appeared in my real-life garden. As more and more cultural creatives *live* from this expanded perspective and join together in creative and innovative projects, the world will be transformed.

For the past few millennia, humanity has been tamed with fear and judgment. Religion has been used as an agent of social control and conformity. Strange and downright crazy beliefs about a wrathful, judgmental God have been ingrained into the human psyche and have convinced us that being spiritual, or worthy of love, means *being good*. That is, doing what others want us to do. Conforming to external rules and social expectations. Putting others' needs before our own. Sacrificing ourselves. Denying ourselves. Trying to prove ourselves worthy or special.

Most of us have internalized a cold, critical, repressive inner voice—the judgmental god within—that tells us how to behave, constantly criticizes us, keeps our noses to the grindstone, and squashes or denies our true feelings. Since emotions and desires are the language of the soul, this is tantamount to soul murder. To the extent that we give our power to this inner judge, we are tamed and locked away. We feel trapped and disempowered. The shutters of our windows are closed. The inevitable result is anxiety, depression, physical illness, dysfunctional relationships, or a pervasive joylessness that we cannot explain.

The great awakening is about shifting into a mature cosmology in which we are not miserable earthworms in search of redemption, nor have we been cruelly abandoned by a distant (or absent) God. Instead, we are divine and creative beings in human form. We are sparks of God in conscious evolution. Paradoxically, this grown-up concept of God and the cosmos has been emerging from the very realm that once rejected God as unnecessary and irrelevant: the world of science. At the cutting edge of physics and biology, a psychoenergetic vision of reality is turning our "common sense" on its head. Within this new cosmology, the universe is an interconnected web of energy-consciousness. It is consciousness that determines the collapse of quantum wave functions, which essentially means that it is our minds that control what happens in our lives, not a judgmental God. No commandments from the sky. No victims of chance, or fate, or karma, or genetic inheritance. Instead, we discover a universe in which love is so unconditional that it says "Yes!" to any desire we have. No questions asked. No need to earn or deserve it. No doubt about whether we are "allowed" to have that desire. The universe has infinite resources, and it can coordinate the higher needs of everyone involved. Ask and it is given. It is as simple as that.

Well, no, it isn't *quite* as simple as that. As the new science shows, energy follows thought. We get what we focus on. Focus on problems, and they get worse. Focus on your fears, and they will eventually manifest. Focus on what you desire, and you pull it toward you. Focus on your limitations, and you cannot move beyond them. What you expect, you get. You prove yourself right, over and over again. You *can* have whatever you want, but you need to send out consistent signals. Your beliefs, desires, and expectations need to be coherent.

What splits our energy, producing incoherent or contradictory signals? *Fear and judgment*—not feeling good enough. Feeling undeserving or unworthy. Feeling guilty about our desires. Fearing that if we have what we want, we are depriving others of what they want. Telling ourselves we should be good and dutiful, or should put others first, or should make the best of what we have. Believing that dreams cannot come true or that we will get our rewards later if we suffer and sacrifice ourselves now. Feeling anxious about making a change. Fearing that life is going to fall apart. A whole host of other negative thoughts stem from the old cosmologies—from the finger-wagging God in the sky, or a belief that the world is a solid and limited place over which we have little control.

Abraham suggested that the basis of life is freedom, the purpose of life is joy, and the result of life is growth.[2] For me, this triad of freedom, joy, and growth is the cornerstone of a grown-up spirituality based upon the new paradigm—a spirituality of unconditional love, in which each of us is a creative spark of an infinite and omnipresent Source of energy-consciousness. Freedom, joy, and growth are the new holy trinity. We are cocreators with the Source energy, which means that we are constantly giving birth to new dreams and desires that we are learning how to manifest in physical reality. We chose to be here in order to have a delicious, glorious, and wonderful time!

The purpose of life is joy. In the New Age movement, it is common to see the purpose of life as spiritual growth, but this readily feeds into the fall/redemption belief that there is something wrong with us that needs fixing. Nor are we here to serve the world. That belief likewise comes from the tame, driven part of us that feels unworthy and wants to justify our existence (or from a part that feels superior). Instead, our purpose is to experience joy. What a relief! Because the basis of life is freedom, how we choose to find joy is entirely up to us. This is a planet of free will. There is no judgment about the choices that we make, now or in the hereafter. Because the universe is aligned with love, joy, and freedom, life tends to go more smoothly and easily if we choose to be happy, rather than trying to be "good" in others' eyes.

This is why I urge people to choose to be happy instead of trying to be good,[3] or as Joseph Campbell put it, "Follow your bliss." If you believe that being spiritual (or loving) means being good, you're in big trouble. After all, you have identified spirituality with the world of the ego—the world of judgment—and will tie yourself in neurotic knots. Trying to be good and virtuous leaves us feeling lonely and disempowered. Whenever we try to be good, we allow ourselves to be controlled by other people, religious rules, duty, obligation, or pressures to conform. We squash or deny any feelings or desires that seem unacceptable to others and twist ourselves into pretzels trying to gain approval or admiration. We cut ourselves off from our emotional/intuitive guidance—*the God within us*—that is constantly guiding us toward the fulfilment of our desires. Life becomes a joyless treadmill. Whenever we dance to someone else's tune, we begin to lose ourselves and crush our unique potential. As in Shakespeare's play *Hamlet*, the Delphic oracle urged, "To thine own self be true." If we remain true to ourselves, honoring our own feelings and desires, we begin to live a life that truly belongs to us, one that becomes more and more joyful, loving, authentic, rich, creative, and meaningful.

Following our bliss honors the divine feminine. As we honor our sensuality, our sexuality, our creativity, our desires, and do whatever makes our hearts sing and our toes curl with delight, we embody our soul. We step into the emergent holistic cosmology, which sees everything as divine and sacred. We melt the boundaries between heaven and Earth. Bliss makes our energy coherent, which, in turn, attracts further joy, and miracles, and synchronicities.

TRANSFORMING RELATIONSHIPS

Whenever we split the world into good and bad, conflict and neurosis are inevitable. Our energy becomes split or divided. Whenever we pride ourselves on saying "no" to a cream cake, working through our lunch break, or pleasing our partner at the expense of our own feelings, we are bowing to the old cosmology. Trying to be good stems from a dualistic way of thinking. It is based upon judgment, or conditional love. It fuels self-righteousness, which means someone is "in the right" and someone else is "in the wrong"—some part of self is right (the judge within), and another part is bad and wrong (our feelings, thoughts, or desires). This inner conflict will be mirrored in conflict with others. Splitting ourselves internally leads to projecting our shadow onto others. This is at the root of wars, terrorism, genocide, racism, sexism, family feuds, religious factions, scapegoating, and most relationship difficulties. As Dr. Carl Jung said, "The psychological rule says that when an inner situation is not made conscious, it happens outside as fate. That is to say, when the individual remains divided and does not become conscious of his inner opposite, the world must perforce act out the conflict and be torn into opposing halves."[4] This dynamic creates a huge proportion of the misery in the world and blocks our natural ability to love with an open heart and to speak honestly without blame or defensiveness.

Gandhi said we need to *be* the change we want to see in the world. Because all energy is interconnected, we are all one at an energy level. What we *are* has far more impact than what we say or do. If we are loving toward ourselves and at peace with ourselves, that helps to spread love and peace in the world. Our energy radiates far beyond our personal lives, like ripples spreading across a great lake. When we are negative or critical, even in the privacy of our own minds, that energy affects the world. When we are loving, joyful, peaceful, creative, and visionary, that, too, affects the world around us.

However, this isn't an excuse to beat ourselves up whenever we are having a bad time, feeling negative, angry, despairing, or fearful. If we react in this way, we slip back into the old cosmology. A crucial part of self-love is accepting whoever we are, wrapping our arms around ourselves, and saying "yes" to whatever is. The paradox is that we cannot change while we are pushing against whatever is. When we push, we are in resistance, which blocks the flow of energy. We have to accept what is in front of us in order to open the door to change, whether it is a negative emotion, a situation we find difficult or painful, other people's behavior, or what is happening in the world. To say "yes" to it, we have to soften our resistance. Our love needs to be unconditional: Tolerance is not love. Approval is not love. Dependency is not love. Love heals and transforms everything, but it has to be wild, or it is not love.

We can truly love only when we *know* we are good and aim to be happy. We cannot love while we are *trying* to be good. The world of judgment traps us in ego—we can only tame others or allow ourselves to be tamed. When we try to please others from a sense of unworthiness, we are trying to "earn" love by sacrificing our true selves; if we emit shame and insecurity, our relationships will reflect that back to us. If we try to be good by identifying with the judgmental god within, we become critical and controlling of others, holding them responsible for our feelings and wanting them to live up to *our* model of who they should be. Either way, love and intimacy are nowhere to be found.

When we aim to be happy, by contrast, we give off vibrations of self-love, self-worth, and appreciation, and our relationships mirror this by becoming deeper, happier, and more authentic. In other words, much of what we have been taught about "what love means"—self-sacrifice, putting others first, being loyal to others at the expense of our own feelings or authenticity, or feeling entitled to have others behave as we wish them to-actually leads us *away* from loving relationships and into the twilight prison of codependency. It leads us toward tame love, which constantly slips into toxic cycles of control and sacrifice, blame and guilt. Tame love splinters our awareness and strangles our potential. It holds us hostage to the old cosmology.

As part of the great awakening, I believe we are in cultural transition from tame love to wild love, from the limited psychology of the ego to the multidimensional awareness of the Self. As a result, personal relationships are currently in chaos. As I see it, the current pattern of marital breakdown and blended families is not necessarily a bad thing. As family therapy reveals, dysfunctional family patterns tend to be unconsciously passed down through the generations. Children grow into adults who mistake for love whatever they received in childhood, since it feels familiar. They then feel "bound by loyalty" to similarly dysfunctional relationships in adulthood—and the chain of pain is passed on. Perhaps the ability to let go of limiting relationships is essential in dissolving our old riverbeds of relating and giving the next generation fresh options and greater freedom. Perhaps it is a crucial part of our spiritual awakening to release our old security blankets to build healthier patterns of relating. If so, we need to stop judging a marriage as a failure if it ends in divorce. After all, some of the *least* successful relationships that I know of have lasted for decades.

Marianne Williamson suggests that marriage often becomes "a prison based on guilt and ownership"[5]—a chilling description of tame love. Tame love traps us in fear, guilt, control, and conformity, making us far smaller than we really are. But we are now outgrowing that old model of love and commitment, with its demands, possessiveness, and feelings of "entitlement," and forging new riverbeds of wild love. We are reaching for a greater love, which allows us to expand our consciousness and express our creative and loving potential—Love that sets us free to be who we truly are.

During the past few years, my own journey has unexpectedly taken me out of a marriage, which—though friendly and companionable—was limiting and two dimensional. It had rocked me to sleep, without my even noticing. Through an experience of wild love, I reawakened my passion, sensuality, and embodiment and reconnected with my emotions. I reclaimed my lost dreams and desires. My journey was immensely tough and painful, yet it released much that for years I had kept under wraps. It smashed through my old defenses. I felt raw, exposed, and vulnerable, as if an old suit of armor had been ripped off, leaving my flesh tender and pink. It changed me profoundly, in ways that I could never have imagined. It helped me to understand how we control and limit ourselves. And it taught me what love, for self, others, and the world, really means.

TRANSFORMING SOCIETY

Behind the scenes, hundreds of thousands of people are shifting in their ways of seeing life and relating to each other. These "quiet shifts" are working their way through all of our social systems and building toward a gentle revolution that will transform everything from health care to education.

It is all too easy to see the problems in the world. However, shifting into the new cosmology requires us to focus on our desires, our visions, our hopes for the future, as opposed to criticizing, analyzing, and explaining what is wrong, or blaming those who are "behaving badly." Energy follows thought. We are dreaming this new world into being, and therefore we need to have peaceful, loving, and visionary dreams.

While we are straddling the old and new paradigms, it is tempting to use the judgmental approach of the old worldview to criticize the current systems. I know—I've been there! I spent 25 years criticizing allopathic medicine and promoting alternative approaches to health care. I criticized patriarchal religion, mainstream education, and other social systems. I learned, however, that we get what we focus on. When we judge systems, people, or ourselves as bad or wrong, we participate in the old world of duality. From that place of self-righteousness, we cannot become an effective agent for social change. Only wild love can set us free.

Eventually, I realized I was focusing on what I saw as "bad and wrong" and perpetuating the split between "us and them." I began to see the others—the doctors, priests, teachers, politicians, lawyers—as good, loving, and well-intentioned. Perhaps some had different priorities than mine or saw the world in a different way, but now I wanted to understand them. I wanted to reach out and connect with them. I began to make peace with the disowned parts of myself, which they mirrored. There is a time when we do need to separate from old systems, when we need to say "no" and walk away. There is a time for establishing clear boundaries, finding our own way, asserting our identity, and clarifying our vision. Then comes the time for peacemaking, reconciliation, and integration.

Some believe that our old social systems are becoming so dysfunctional that they will simply collapse, that we are heading toward chaos and disintegration as we approach 2012—then a whole flock of phoenixes will rise from their ashes. My own vision is for a peaceful and gentle revolution. I believe that we can integrate new perspectives into the old structures, thus bringing a loving integration to all our warring selves and initiating change from within.

For example, I recently cofounded a project to bring broader spirituality and deeper ecumenism into my local church. We offer additional meetings and services that are creation-centered, joyful, and celebratory, while the traditional Sunday morning services continue to meet the needs of those who are comfortable and familiar with the old fall/redemption theology. Our aim is to be integrative and to honor the needs of everyone involved. The new project offers a positive way forward for a small church, which, amid widespread spiritual hunger, was failing to attract more than a handful of people for its congregation. Instead of labeling the old approach as bad or wrong, we recognized that it was meeting some people's needs and had its own strengths. Introducing complementary services helped to embrace a clear vision for a bright new future.

Loving acceptance is a key part of this new approach. Parents do not punish a toddler for being unable to run and climb, or shame the toddler for wobbling on his or her feet. Nor should we criticize those parts of ourselves—mirrored in the outside world—that are still clinging to the old paradigm. Those old habits we have developed have been well trained over several millennia. They serve the belief that striving to be "good," or self-righteous, is the only way to remain safe and loved. Perhaps they are holding wisdom and gifts for us, if only we listen with an open heart. Until we love and accept every part of these behaviors, old and new, they will continue to resist change and cling to the raft of fall/redemption theology. This includes all the ways in which that theology plays out in society, such as allopathic medicine, left-brained education, the blaming and victimhood of the legal and penal systems, power-based hierarchies in business, child abuse, disempowerment of women, or ecological problems. Somehow we have to maintain a delicate balance between being aware of how the old cosmology controls, limits, and distorts us, while not pushing against it in a way that increases resistance to change.

The only solution is love: not tame, conditional love that says, "I will love you if only . . ." or "I will love you when . . .", but unconditional love that says, "I see you. I hear you. I love you." Wild love. Love that says "yes" to everything; love that is inclusive and embracing rather than excluding or rejecting. Wild love heals the old wounds and divisions. Until we reach out with love, compassion, and understanding toward those who might seem stuck in old thinking, or even acting out in destructive ways, we are playing the same old games of goodies-and-baddies. We are caught up in fear or judgment. We are living in the old world of duality.

At an inner level, reaching out with love symbolizes an integration of our inner selves. This makes it feel deeply nourishing. We cannot reject or exclude any aspect of the self without paying a price, so we cannot feel truly at peace until we have made peace with how the world is right now and with those we might have judged or wanted to push away. There might be room for change and improvement, but evolution and growth are the very nature of life. The world will never stand still, but in any system, there are conservative forces that strive to maintain the status quo. Perhaps those conservative parts of ourselves are serving a useful function, protecting us from too rapid a change or helping us to clarify our true desires. Each of us is a unique piece in this giant cosmic jigsaw puzzle, and every part is needed to make the puzzle whole.

TOWARD 2012

Personally, I don't believe that anything dramatic will unfold in December 2012 (although those who fear Armageddon might create their own private doomsday). Yet, when we look back, we will see a sweeping change in human consciousness over the past decade, as if we had collectively awakened from a dream.

As we move toward the great shift, countless people are letting go of old habits, patterns, relationships, and situations, often being apparently *forced* to release whatever holds them in the past, or keeps them stuck. We can no longer pretend that we are happy if we are not. We have to face up to whatever is not working or falls ·short of our dreams and desires. This means we must become accustomed to rapid change, personally and globally. If we can learn to ride this tidal wave of change, then our global awakening will be relatively easy and even exhilarating. The more we resist change or close down, the tougher it gets. We have to be willing to let go, to make unexpected changes in our lives. We have to be willing to open our hearts, to follow our bliss, to trust and go with the flow. We also have to be willing to put down roots and make commitments when that is where our hearts lead us. As Paulo Coehlo suggests, freedom is not the absence of commitment, but the ability to commit to whatever is right *for you.*[6] *That* kind of commitment—unlike one that comes from duty or obligation or promises—always feels ·joyful and liberating and expansive. If it feels heavy or restrictive, it invariably means we are "trying to be good" and so are splitting or dividing our energy.

A healthful cosmology can transform our everyday lives. Instead of seeing life as a harsh training school for wayward souls, or a karmic wheel from which we might eventually escape (if we are good or lucky enough), or merely as a statistical accident with no inherent meaning or purpose, we instead see life as a wondrous gift. We are not here to be good or perfect. We are not here to prove ourselves worthy. We are not here to serve others (at our own expense) or to save the world. We do not have to earn or deserve love. We do not have to "behave well" or conform to external rules and expectations. In a loving universe, we can relax. We are safe. We are worthy. We are loved without condition. We are cosmic voyagers on a magnificent adventure in physical reality, and—as creative sparks of the divine—we can have, do, or be anything we wish. No limits. No strings attached. We can create our own heaven on Earth. And the key to doing so is unconditional love—for self, others, and the world.

Wild love does not separate the world into good and bad, right and wrong, safe and dangerous. It says a profound "yes" to life. It affirms that "everything within you and me is good." It is fearless relating. It dwells in a blameless universe. It is love that gives without needing to receive. It is love without needs, demands, or expectations. It is love that inspires and expands us. It is love that sets us free. It transforms us from flickering lightbulbs into luminous laser beams. Wild love allows us to focus our energy and intention in ways that can change our lives dramatically. *Everything we experience depends on whether our love—for self, others, and the world—is tame or wild.*

When we love wildly, we take responsibility for our own experience instead of blaming others or trying to control them. We know that nothing and no one "out there" is responsible for what we feel or experience. Everything mirrors what we hold in our own consciousness. Nothing can block the fulfillment of our desires except us—our resistance, our fear, our judgment. Instead of focusing on what is bad or wrong, we learn to embrace ourselves, others, and the world. We appreciate all that is wonderful in our lives and in other people. Instead of pushing against what we do *not* want, we focus on what we *do* want. We accept what is and reach toward what might be. We change our world from the inside out. Then, we reconnect with who we really are. We are awake. We are alive. Our wildest dreams can come true. And, almost without noticing it, the boundaries between heaven and Earth have melted away.

The Bible Code

LAWRENCE E. JOSEPH

*Forecasts for 2012 have largely been associated with cultural
creatives and independent researchers. In his book* Apocalypse
2012, *Lawrence E. Joseph reviews a host of reasons 2012
may be a dramatic, possibly destructive time, and also takes
a look at some uniquely Christian perspectives relating to 2012.
In the following excerpt from his book, Joseph tackles 2012 in
the context of* The Bible Code *and Christian fundamentalists.*

The Bible tells us that God will annihilate the Earth in 2012.

This is the conclusion reached in *The Bible Code*, Michael
Drosnin's global bestseller that plausibly purports to have
decoded a secret, divine code embedded in the text of the Bible.
The diamond-hard basis of this claim is a scholarly article entitled "Equidistant Letter Sequences in the Book of Genesis,"
written by three Israeli mathematicians, Doron Witztum,
Yoav Rosenberg, and . . . Elijah, in the person of Eliyahu Rips
(Eliyahu is the transliteration of the proper Hebrew spelling of
the fiery prophet's name), and published in *Statistical Science*.
This remarkable piece of statistical analysis verifies an observation first made by H. M. D. Weissmandel, a rabbi in Prague,
that if he "skipped 50 letters, and then another 50, and then
another 50, the word 'Torah' was spelled out at the beginning

of the book of Genesis." The same skip sequence yielded the word "Torah" in Exodus, Numbers, and Deuteronomy, the second, fourth, and fifth books of Moses. (For some reason, this procedure does not hold true for Leviticus, the third book of Moses, which spells out the rules for priestly comportment.)

That discovery piqued the researchers' curiosity to see what else might be encoded. The task was daunting: Isaac Newton taught himself Hebrew and spent decades searching for the code he felt sure was embedded in the Bible. Newton, perhaps the greatest scientific mind in history, came up with nothing. That's because he didn't have a computer. The three Israeli mathematicians input the Book of Genesis, in the original Hebrew characters, letter by letter, with no spaces, no punctuation, just as biblical texts were originally scribed. In essence, they laid out Genesis as a huge acrostic and then searched for words to circle—vertically, horizontally, and diagonally. With the help of the computer, they examined this acrostic for words not just composed of adjacent letters but also of letters separated by a given number of spaces, just as Rabbi Weissmandel had originally done in finding the word "Torah." Using the same equidistant-letter, skip-code approach, computer analysis yielded the names of 66 legendary rabbis, virtually all of whom lived many centuries or millennia after Genesis was written. In each case the names were close to, or intersected by, the rabbis' birth and death dates and cities of residence.

Certainly no mortal could know, and also surreptitiously encode, the names of such venerable holy men so many centuries into the future. The implication is clear: There is a secret code embedded in the Bible, placed there by God. Eliyahu Rips, Drosnin's chief collaborator, explained the seeming impossibility by citing one of the rabbis discovered in the Bible, yet another Elijah, the famed eighteenth-century sage Rabbi

Eliyahu of Vilna: "The rule is that all that was, is, and will be unto the end of time is included in the Torah, from the first word to the last word. And not merely in a general sense, but as to the details of every species and each one individually, and details of details of everything that happened to him from the day of his birth until his end."

It's as though the U.S. Constitution were found to contain the names of 66 future presidents, intersected by or adjacent to their home states and the dates they were elected. Or if the 1965 annual edition of *Sports Almanac* contained the names of the next 66 Super Bowl winners, with the scores of the games.

The Israeli mathematicians' rigorous statistical analysis concluded that there is virtually zero probability, 1 in 50,000, of this all being a chance occurrence. Not surprisingly, their extraordinary claims came under a barrage of attacks. In the decade since this paper was published, a number of statisticians and mathematicians, including experts from the National Security Agency in the United States, have challenged their findings by disputing their methodology and by running comparable tests on two other original Hebrew texts as well as the Hebrew translation of *War and Peace*. To the best of my knowledge, no such test has been run on the Quran. But no one thus far has laid a glove on the Israeli mathematicians. Indeed, some of those who set out to refute the existence of what has come to be known as the Bible code are now among its most ardent supporters.

Drosnin, a journalist, began digging into the Bible codes for clues about the future. His most famous discovery is that the name Yitzhak Rabin is physically crossed by the phrase "assassin that will assassinate." Further deciphering indicated a place, Tel Aviv, and a date, 1995, which at that point was still in the future, of course prompting Drosnin to make every effort to warn Rabin, but to no avail. After Rabin's tragic murder, Amir, the name of the right-wing assassin, was found to be encoded nearby.

What's next? Drosnin logically wanted to know. Out poured any number of observations and predictions, mostly concerning the Middle East. Drosnin tends, as so many of us with Semitic heritage do, to equate the outcome of that region's interminable drama with the fate of the world. I have never once heard of anyone from the Southern Hemisphere imply that our collective fate hangs on the outcome of their regional disputes. Although Drosnin has been duly criticized for overinterpreting the Bible code, an impressive proportion of his predictions have since panned out, including a prediction that a judge, in this case the U.S. Supreme Court, would rule against Al Gore and in favor of George W. Bush in the matter of the 2000 presidential election.

Let's all pray that Drosnin's hot streak cools. Because according to *The Bible Code*, comets are expected to pound the Earth in 2010 and also 2012 (5772 in the Hebrew calendar), at which point the "Earth annihilated" prediction also comes into play. True, his analysis also unearthed the phrases "It will be crumbled, driven out, I will tear it to pieces" near the 2012 comet, though that could be a mixed blessing, causing the Earth to suffer multiple major impacts, potentially more damaging than one big blast.

The Bible Code provides the most profound scientific evidence yet that the Bible was divinely inspired. The work of the Israeli mathematicians Rips, Witztum, and Rosenberg has withstood all scientific challenges thus far. The good news is that the book upon which so much of the world's religious faith is based has received unprecedented mathematical substantiation. The bad news, of course, is how the Bible story ends.

Then I saw coming from the mouth of the dragon, the mouth of the beast, and the mouth of the false prophet, three foul spirits like frogs. These spirits were devils, with the power to work miracles. They were sent out to muster all the kings of the world for the great day of battle of God the sovereign Lord. (That is the day when I come like a thief! Happy the man who stays awake and keeps on his clothes, so that he will not have to go naked and ashamed for all to see!) So they assembled the kings at the place called in Hebrew Armageddon.

—Revelation 16:13–16

Some say that from atop Armageddon, the fabled hill that looks out over the Megiddo plain in Israel, you can see the end of Time, because that's where the Battle to End All Battles will be fought. (In Hebrew, *har* means "hill," and Megiddo becomes "mageddon.") Prophesied in Revelation to be the site of the final clash between Good and Evil, that is, between those who accept Jesus Christ and those who do not, Armageddon looks down on a 200-mile-long valley that will one day be filled with corpses, two to three billion of them, according to some scholars' extrapolations. It's supposed to be quite a majestic view. But you couldn't prove it by me. I will never set foot on Armageddon. And I hope you never set foot there either, or if you already have, I hope you never return.

Armageddon refers to the great, consuming war to be fought among the peoples of the Earth. Apocalypse is the natural/supernatural cataclysm expected to come after Armageddon. I am opposed to all of it, no matter how "enlightened" the aftermath will supposedly be. (If it all turns out to have been worth it, then I will crawl out of my earthen bunker and admit my mistake.) Now, trying to oppose global catastrophes such as supervolcano eruptions or comet impacts would seem about as effective as trying to oppose the law of gravity. But Armageddon is different. Of all the potential cataclysms, Armageddon is the only one for which significant numbers of Muslims, Christians, and Jews actually hope, pray, and scheme. And it's the one end-times prophecy that we actually might have the power to prevent, or fulfill.

Karl Marx observed that when a theory grips the masses, it becomes a material force; sadly, Marx's theories did just that for more than a century. The doctrine of Armageddon has gripped several small but extremely motivated and influential groups in the United States, Israel, and the Muslim world, and that doctrine is rapidly becoming a powerful, perhaps unstoppable, force in global politics.

"While most Jews, most Christians, most Muslims, most everybody abhors and eschews the harness of fundamentalist thinking, history is not driven by most of us. . . . As a rule, majorities are ruled. It's the fanatic few, at whom we may laugh one day and cower before the next, who are history's engine. It's a minority of single-minded maniacs who can take a holy place and make an unholy mess," observes Jeff Wells, a blogger for the webzine Rigorous Intuition.

Far more disturbing than its hold over a few zealots is Armageddon's powerful crossover appeal. *The Late Great Planet Earth*, by Hal Lindsey, which predicted that the great Armageddon battle would come in 1988 or thereabouts, was the best-selling nonfiction book of the 1970s. Israeli tour operators have seen their business double and double again as impassioned Christians from the United States, Europe, and elsewhere flock to the region. Indeed, a recent survey conducted by the Israeli ministry of tourism indicates that of the nation's two million visitors each year, more than half are Christian, and more than half of those identify themselves as Evangelicals.

Evangelical Christians are the group most eager to precipitate Armageddon, looking forward to the Rapture, the exalted moment when, before the battle begins, true and faithful Christians are literally lifted up into the air, into the heavens, to join God. No doubt this would be exhilarating. From the safety and comfort of Heaven, one would have the opportunity to look down upon the Earth and watch the battle between two warring groups: Christians who, due to imperfections in their faith or because of special warrior destiny, were not subsumed in the Rapture; and followers of the Antichrist, a charismatic false Messiah, whose followers include secular humanists, pagans, Hindus, and Buddhists, as well as Muslims, Jews, and insufficiently committed Christians. A large proportion of Jews are expected, in Evangelical theology, to convert to Christianity and thus fight on the righteous side of the Armageddon battle. Those who decline Jesus will, along with all other naysayers, explode.

The more people who go to Armageddon, the more mystique that hill gains, and the more likely that some incident, spontaneous or staged, will ignite a tragic war. Wave upon wave of pilgrims will soon besiege the new Christian theme park being built nearby, on a 125-acre stretch along the Sea of Galilee, where Jesus Christ is reported to have walked on water. The $50 million project is being developed by a partnership between the Israeli government and American Evangelical groups. According to a spokesman for the National Association of Evangelicals, a 30-million-member group spearheading the project, the Galilee World Heritage Park should open in late 2011 or early 2012.

Things are just breaking the right way for the Armageddonists these days. What may be the oldest Christian church in the world was accidentally unearthed in Megiddo in late 2005 by Ramil Razilo, a Muslim prisoner serving a two-year term for traffic offenses. Razilo was part of a crew of inmates helping to construct a new facility to detain and interrogate Palestinians. Armageddon Church, as it is now known, dates back to the third or fourth century, a time when Christian rituals were still conducted in secret. At the center of the 24-square-foot mosaic on the floor is a circle containing two fish. The fish is an ancient Christian symbol; the spelling of the Greek word for fish makes an acrostic for the name Jesus Christ. Early Christians greeted each other by making the sign of the fish, which also alluded to the apostle Peter, a fisherman, who went on to become a "fisher of men." Peter's name, which means "rock," was an allegory for the rock upon which the Christian church was built, most famously, St. Peter's Basilica at the Vatican in Rome.

Although there is no specific biblical foretelling, this discovery is already being hailed as yet another sign that the end is near. Estimated restoration date of Armageddon Church in Megiddo: 2010–2012.

An Awakening World

LLEWELLYN VAUGHAN-LEE, PH.D.

The year 2012 is often called the beginning of the Golden Age. In the following essay, Sufi teacher Llewellyn Vaughan-Lee describes the possibility of a global awakening into an era of oneness, in which an awareness of the divine will return to everyday life. But what is important is how we participate at this present moment and how we help the world awaken.

May we be those who shall renew this existence.

—Zarathustra

A TIME OF TRANSITION

Many have prophesied that the year 2012 will be one of tremendous change. In the Mayan calendar, 2012 marks the end of time; other prophecies see it as the beginning of a new Golden Age. As we approach that year now, our world is indeed undergoing accelerating change—does it portend an end, or a beginning, or both?

We have entered a time of global crisis. We live surrounded, almost suffocated, by the debris of our dying civilization. The last centuries' belief in dualism and separation with its focus on rational thought brought great scientific and material progress, but it also created a life-denying split between spirit and matter.

It banished God to heaven and stripped the Earth of its sacredness. Now, all reverence for our own bodies and the body of the Earth lost, we are systematically plundering and polluting our world, destroying the very ecosystem we depend upon for survival. Steeped in materialism, we have forgotten why we are here. We have created a physical and spiritual wasteland, and our own souls and the soul of the world are starving.

There is a tradition that this era of separation will be followed by an era of oneness, in which the simple awareness of divine presence will be given back to humanity, and the divine will once again return to everyday life. If we look closely, we can see amid the destruction the hints of something new, the seeds of a global consciousness, a deep awareness that we are all one people, part of an organically interrelated living system. And within that awareness, another consciousness is awakening, a primal knowing of the oneness that belongs to life and is a direct expression of the divine.

There is also another tradition that says that just as an individual may go through a spiritual transformation, so can the world. There is an ancient promise that the world—as a living spiritual being—can awaken and that its heart may open and begin to sing. What this might mean we cannot even imagine. There are hints in our ancestral memories that once the world had been awake and sang the sacred song of creation, when everything had been fully alive and magical and a part of this song. Maybe this was the Golden Age, which can now be reborn.

These are prophecies, possibilities, portents, promises. But what every spiritual teaching tells us is that only the present moment is real. Only in the *now* are we alive and do we have access to the divine that is within us and within life. What matters is how we use the moment. Do we remain in this self-destructive, narcissistic dream, amusing ourselves with the toys of triviality? If we dare awaken to the present, we may hear a call from the depths of creation, crying out for us to work with it, to help it in this time of transition, to midwife the future, to help the world to awaken.

A NEW LIGHT

Life is one, has always been one. The world is a single, living, organic wholeness, and everything in creation is a part of it, as vitally and inseparably related to the whole of life as an individual cell or organ is related to the larger organism of which it is a part. Life is alive with its own intelligence and purpose. That purpose, which is our own purpose because we are not separate from life, is one of revelation. In its multitude of forms, creation reveals the hidden face of God—"I was a hidden treasure and I longed to be known, and so I created the world."[1] Through witnessing the creation, we can come to know the Creator.

But we have forgotten this. Our present world bears witness not to the divine but to our own hubris, our greed, our delusions of power. In our obsessive desire for material comforts, for more and more *things*, we have plundered the Earth as if it were a mere commodity to be exploited, and we have desecrated our world. We live as if life were something apart from ourselves—something we can master and control—rather than part of our very being. We have lost all sight of life's real nature and purpose and our relationship to this purpose. And life has allowed this to happen. Until now, life has, for the most part, remained receptive, waiting, watching us follow our desires, and saying very little, even as it has felt the sorrow of a people who deny its divine nature.

Now, something deep within life is changing; an era is ending, and at the core of creation, something is coming alive in a new way. A light at the center of the world that has been dormant for millennia has been rekindled. This is the light of life itself, waking up, remembering its own real nature and divine purpose. With this awakening, the living being that is our world is undergoing a transformation in its very essence.

The light of this awakening is extremely dynamic. It is metamorphosing very quickly, bringing into its orbit other energies, other forces within creation—some awake, some still dormant—and is setting energies in motion that are going to change our world. The awakening of the light at the center of the world carries the potential for a whole new revelation—the possibility of a new way of living and being and relating to one another and to life—in which creation's divine oneness is revealed and celebrated. At this moment, we stand on the edge of this new stage in the evolution of life and consciousness, a new paradigm for the world.

There is a growing awareness of a shift in consciousness toward global oneness. With regard to ecology, we have been forced to recognize that we are one planet, one interconnected living ecosystem. Internet and cell phone technology has given us the tools of global communication, and commerce and finance also have a global dimension. Few, however, have recognized the depths of this change, the signs that the world is awakening. Most are too busy, too caught up in their own affairs to notice something so new. Those who are fixed on the image of an Earth that has no soul will not recognize its new light, will not allow it into their lives. It would be too threatening. Yet this awakening light is beginning to affect human consciousness. It is sending messages of hope into a collective despair, a collective soulless-ness. And it is attracting those who want to work with it.

This light needs our human consciousness to work with it, so that it can become fully alive and realize its potential. It needs us to be awake at a moment of awakening, to be alive at this moment of rebirth. And it calls to each of us in our own way, reminding us of an ancient pledge to witness and then participate in what is being born. It is our remembrance and our awakening. It is the light within us, our own knowing that we are one. Yet, it is also something other, something that is being given to this world.

Stepping into the arena of this light, we become a part of what is being born: we contribute our individual light to this work, our consciousness to this awakening. We welcome divine oneness and divine presence back into the world. And we also hold back the forces of darkness, the demands of greed, fear, and power that would otherwise try to abort this new birth, try to ensure that nothing can be born. In our prayers and our awareness we protect this new birth, this awakening of the light of the world. This is a simple commitment from our soul to the soul of the world, an acknowledgement of its need for us and our need for what will be born.

To be present with this light, we need to leave behind many of the dramas and conflicts that belong to our egos' versions of ourselves and the world. We need to relate to it with our own light, the pure awareness within us that belongs to a higher dimension of our being and to our own instinctual awareness of the oneness of life. When we do, we will recognize that we are not separate from life, and the light of the world lives within us. We *are* the light of the world; we carry the consciousness of life's highest purpose within the light of our true being. It is present in our every breath. It is our remembrance of the divine, our essential nature breathing the light and life of this world. The awakening of the light of the world is our own as we wake up to our own essential nature and purpose, to the primal knowing that we are one.

But it is also something completely new, a new energy and consciousness, that is being given to this world at this moment in its evolution. And when we bring the light of our consciousness to the light of what is coming into being, we become a partner in its creation. This is real cocreation: the interaction of light with light, our individual light and the world's light, the dynamic interplay of the one light of His love.

Our true light comes from the divine spark within us. It carries the deepest purpose of our life, which belongs to the divine purpose of creation. We are born with this spark; it is quite visible in children, in their eyes, their laughter and joy. But as we grow older, it "fades into the common light of day."[2] Then, we must reclaim and reignite it through the sometimes painful process of remembrance and reawakening that spiritual practices and inner work provide. This is the real human consciousness, which belongs to the consciousness of the Self. And it is far more powerful than we realize. We might have some sense that our consciousness can shape our own lives, perhaps we have observed the way a simple change in attitude can alter our circumstances, but we have no idea of its power as a catalytic and directing force in all of creation.

Now, we are being asked to wake up to our light and assume our real cocreative role in the evolution of the world. The light of the world is waking up, and our consciousness is needed to help bring its hidden meaning into the consciousness of the world and of humanity. We can help the world come alive again. And we can help to shape the next stage in its evolution.

How do we participate in this work? We start by simply returning our attention to life itself, away from all the distractions and false images of life we have created for ourselves. Simple practices of awareness can help. The basic practice of awareness of breath turns our attention back to what is essential. Simple acts of loving kindness toward others reconnect us with the real web of life, the genuine interconnections between people. Learning to walk in a sacred manner—for example, the Native American practice of walking with the awareness of placing the foot on the sacred Earth—can bring our attention back to the sacred nature of the world. The "Practice of the Presence of God" described by Brother Lawrence,[3] in which one does one's daily actions in the awareness of God's presence, reaffirms the divine in daily life.

Once we acknowledge and then recognize the world as a divine living being in which we are present every moment of every day, our own divine light will be present in life. It is only in the simplicity of the moment that we can participate, that our light can be present. Without the conscious awareness of being part of creation as a living being, we stand on the periphery, unconscious spectators of our own destiny and the destiny of the world. Taking responsibility for our planet does not mean just an ecological awareness, but a reverence for Earth's divine nature and a recognition that the light awakening in the world needs our presence and participation. It needs our divine light. And because our light belongs to our real consciousness, it is through our conscious awareness that our light interacts with the light within the world.

Every moment the light of the world is changing, evolving. It is the dynamic core of life that carries within it the secret of creation, the word of God made manifest. Just as our light carries the real meaning of our life, the purpose of our soul, and the knowing of how to live this purpose, so does the light within the world know the world's real meaning, its soul's purpose, and what is needed to live it. This light is ancient and yet new, a continual response to the divine moment. It carries the real history of the world and the real knowing of the purpose of this wonder in which we live. At this time in the world's evolution, this wonder is becoming more conscious, and it needs our light to bring its hidden meaning into the consciousness of the world and of humanity. We are needed to help the world come alive.

IMAGES OF LIFE

Through the simple acknowledgment of life's divine nature as we go about our ordinary lives, we open ourselves and the world around us to dimensions within life that have been veiled from us by our forgetfulness, our addiction to materialism. We have all but forgotten that life is a rich and mysterious coming together of many worlds. We have lost sight of what the ancient priestesses and shamans knew, that the forms of our visible world have their roots in unseen dimensions, and that it is in these unseen dimensions that the primal energies of life lie. In our forgetting, we have lost the wholeness of life, and we have cut ourselves off from the real forces that shape our world. When we are consciously present in life, free from demands or agendas, when we allow life to unfold according to its own inner principles, we open up a doorway again between the worlds. Within our consciousness, the outer and inner worlds, the visible and the unseen, can come together and speak to each other, and our world can become whole again.

The reconnecting of the worlds is central to the reawakening of life, for it is in the unmanifest symbolic realm that the primal energies of life constellate into the archetypal patterns that shape our lives. When we allow the meeting of the worlds within our consciousness, we regain access to this realm and the powers at work there. We ourselves become the place where the worlds meet, where the primal powers of life, the hidden archetypes, emerge into form. This is the place where the patterns and images that shape the collective life of an era come into being. Standing at that meeting point in ourselves, we can use our awareness to influence their shape and direction.

With the awakening of the world's light, the inner structure of life itself is changing, releasing primal energies of creation that have not been active for millennia. As they again come into play, they give rise to a new range of possibilities, new patterns through which life can flow and the divine can reveal itself in the world. These are the archetypal patterns that will determine the fundamental structures of the next era, the riverbeds through which the waters of life will flow. At this time, the primal energies are still in flux. By directing our consciousness—that awareness in us that is attuned to life's and our own highest purpose—toward them, we can influence the patterns they will ultimately form; we can help determine the structures that will be born from them, the archetypal images that will shape life in the coming era.

This is a time of enormous potential, but also of frightening uncertainty, as the old patterns of life dissolve and the new ones have yet to form, and potent new energies are released. The light at the core of creation is *life coming alive*. It has the violence and power that belong to life. Despite all our illusions of control, we cannot control or contain it—nor should we: its primal potency is needed to revitalize the world. We are not here to control life; we are here to witness and revere the divine as it manifests in a new way through the natural forces of life. We need to bring the light of our consciousness to the awakening light within life so that these primal energies can be creatively channeled, so that they do not unleash too much destruction and chaos. Through our consciousness, we can help direct these energies so that they flow into images of life that support the whole of life—the visible and the unseen worlds, physical life and the life of the soul—so that they reflect life's oneness, its real nature and divine purpose.

Part of this work is simply to be awake, alert to the images emerging both inside us and in the world around us. Some images have already appeared in the outer world—for example the World Wide Web (which has striking similarities to the ancient Hindu image of Indra's net)[4] and the image the first astronauts gave us of the world seen from space, a single beautiful sphere without borders or divisions—as well as in the archetypal world, where the number eight, for example, is emerging as an important symbol of the two worlds coming together.

But unless we recognize these symbols for what they are and welcome them into our consciousness, they will not achieve their full potential. They are much more than just pictures of our world to help us understand and navigate it. They come from the symbolic depths of life, and they carry life's numinous and primal energy. They have the power to create forms, shape events, and reconfigure the collective consciousness. They act as intermediaries between the visible, physical world and the inner realm of light. When we recognize them as such, they align our consciousness with life's inner unfolding—for example, the figure eight or infinity symbol aligns us with the way the energy will flow effortlessly between the worlds in the coming era; the World Wide Web aligns us with the dynamically interconnected nature of oneness—and allows us to participate consciously in the changes taking place at the heart of life. When we bring the divine spark of our higher consciousness to these new images of life's awakening, our spark speaks to the light at the core of creation and helps the images come alive in this world with the full potential of life's highest consciousness and purpose.

But we will have to be awake to the real nature of what is awakening in order to recognize them. The images of the coming era will reflect life's oneness, its aliveness; they could emerge anywhere and will take unexpected forms. The Internet is a striking example. The Internet is a powerful image of the changes happening at the core of creation, yet its deeper meaning and power have been veiled, partly by its very ubiquity. Carrying the energy of life's aliveness, it has quickly become so much a part of the fabric of our lives that we do not see its deeper symbolic dimension and the ways in which it is working to restructure our world. It is still waiting to be recognized as one of the archetypal images of our future—a dynamic image of life's awakening to itself as a single living organism, rooted in the symbolic realm and carrying the real energy of life itself. The Internet not only reflects but also enacts life's living oneness, its capacity to evolve and adapt, to quickly and continually reconfigure itself in an endlessly shifting web of connections.

We need to recognize its sacred dimension, its symbolic nature, if it is to manifest its full potential. Otherwise its higher purpose will be overshadowed by the ego-centered interests and desires the Internet currently serves, and the energy will flow in at a lower level. All the emerging images of life need us to open to their symbolic dimension, so we can reflect back to them the light of life's divine purpose. As the primal energies flow into form, our recognition can help life reconfigure into the patterns, the archetypal images, that will allow life to flourish again as a conscious expression and revelation of the divine and its all-embracing oneness.

There are no books, schools, or traditions that will tell us exactly how to do this work. Only in the meeting of our consciousness with the awakening consciousness within life—our own awakened consciousness as we bring it to life in each moment—will we find the information we need. For at its core, life knows everything it needs to sustain itself; it knows how to evolve and how to recreate itself anew, and because it is a continual response to the divine moment, it knows precisely what is needed in each moment. It knows where our attention is needed and how to direct it in the most beneficial way. It knows exactly what is required to bring our afflicted planet back into balance, to heal the Earth and the soul of humanity that has so painfully lost its way. And as life reawakens, this information will once again become available to us. Life is ready to teach us. Through a dialogue of light with light, the intelligence within life can directly communicate with our intelligence, and it will show us what we need to know.

What life knows about itself reaches far beyond the limits of our current understanding and the scope of our present sciences. When we bring the light of our consciousness into dialogue with the world's light, we will open ourselves to new realms of knowledge and understanding, to a whole new science of life and light as yet undreamed, to a whole new range of human possibilities. When we open ourselves fully to life as it really is, we will experience again that life is alive and that it can communicate with us directly and interact with us in many different dimensions—not just the physical dimension we now take to be the world, but also in the symbolic realms and the realm of pure consciousness and light—dimensions of magic, wonder, and awe that have not been manifest in our world for a long time. When we once again allow life to express its profound intelligence, its playfulness and humor, its delight and passion, we will find ourselves present in a world that is more dynamic and fluid and full of possibilities in each moment than we now believe possible.

A CHOICE

There are primal changes taking place within life, and they are farther-reaching than we dare to imagine. One thing is certain: the world as we know it is ending. Our drama of greed and materialism, our hubristic belief that we are the lords of the world, which we can exploit at will, has created a nightmare that is draining the very lifeblood of the planet. But in the midst of this dying dream, another dream is being born, one based upon the consciousness of oneness, the knowing that we are one living ecosystem, one global community. And within this dream of oneness are signs that the world is awakening.

Perhaps the prophecies of the year 2012 point to this possibility, a moment in cosmic time when, with an outpouring of energy, the world will awaken to its divine nature and throw off the debris of materialism. Those identified with the dying dream would experience this as a catalysm, a global disaster. But others might recognize it as what we have unknowingly been waiting for, a new Golden Age in which we can return to the simplicity and joy that belong to life in its essence, life as it really is—a time when we will no longer need to distract ourselves with our toys and addictions, because the simple wonder and joy of being alive will nourish us and our souls; when the song of the soul of the world will sing to us and to all of creation, and we will discover that the magic inherent in life is healing and beneficial, and that in its wisdom and oneness the world knows how to support itself and its inhabitants.

But what matters is how we respond to the present moment: whether we are prepared to be fully alive at this time of change. Are we prepared to help life to awaken, or are we too frightened of what we might lose, too caught up in the shadowlands of our attachments and collective beliefs? If we are free enough to welcome a future that honors the sacred nature of life and its transformative potential, we will discover we have a central part to play. Our light is a part of the light of the world awakening to its divine nature. Working with it, we can become cocreators of the next era. This is the choice we are being given: not to wait for the future but to help bring it into being.

2012

THE START OF A NEW ERA

JANOSH

As 2012 approaches, many claim there are messages here on Earth that hold the keys to transformative spirituality and consciousness. Crop circles, some claim, may contain messages about the coming era of 2012. In the following essay, Janosh, an internationally acclaimed artist, decodes some of the messages that he believes will help us prepare for the coming era. He answers questions like, What do the crop circles really mean? Who makes the majority of complex, inexplicable crop circles? What do crop circles have to do with 2012?

The year 2012 has long been of special interest to many. This date is fast approaching, and, according to the Maya, we will leave the world of the fourth Sun and enter the world of the fifth Sun. Mayan prophecies indicate this date will mark a time for change, harmony, and rebirth. Today, 2012 represents a transition into a new era, if we only pay attention and open up to it in both visible and invisible ways. By doing so, we will understand what is happening and why it is happening. We will be able to stimulate and awaken our subconscious and participate in creating the new reality that is to come. There are many ways to facilitate this change. For this period to

come, I have found that I can help others—through my art—
transition their subconscious. I will share this with you, but
first, let's talk about this new era that is upon us.

PERIOD OF TRANSITION

Many have questions as to what the transition from the fourth
world into this fifth world will mean. Guatemalan researcher,
historian, and anthropologist Carlos Barrios, who spent 25 years
studying the Mayan calendar system, speaks of the period leading
up to 2012: "We are no longer in the World of the Fourth Sun,
but we are not yet in the World of the Fifth Sun. This is the time
in between, the time of transition. As we pass through transition
there is a colossal, global convergence of environmental destruction, social chaos, war, and ongoing Earth changes. All this was
foreseen via the simple, spiral mathematics of the Mayan calendars. Everything will change." Barrios, among many other experts,
predicts that December 21, 2012, will be the start of a new era,
resulting from the solar meridian crossing the galactic equator and
the Earth aligning itself with the center of the galaxy.

RETURN OF CONSCIOUSNESS

According to Guatamalan Maya Elders of the Eagle Clan, this
fifth cycle will be one of wisdom, harmony, peace, love, consciousness, and the return of natural order. They also agree that
it will not be the end of the world as some may fear.

> *The first cycle was a feminine energy and its element was fire.*
> *The second cycle was of masculine energy and its element was*
> *earth. The third cycle was a feminine energy and its element was air.*
> *The fourth cycle is a masculine energy and its element is water. The*
> *fifth cycle will be a fusion of both feminine and masculine energies.*
> *It will be a transition where there won't be any more confrontations*

between the polarities. It will bring balance, and there won't be
hierarchy of one over the other. Both energies will support each other.
It is why this period is called one of harmony, the kingdom of love
and the return of consciousness. Its element will be the ether.

Based upon the Maya's knowledge of the universe (a knowledge that is still stored inside our subconscious minds yet forgotten and no longer used by Westerners today), we should not fear what is about to happen in a few years time. Instead, we should trust their ancient wisdom and look forward to this transition into a new world, a new dimension.

Many are curious to know what this return of consciousness means and how it will become apparent to us. From the Mayan predictions that date back centuries, it seems a rather big leap to our everyday lives in the here and now. Yet, the same information on this aforementioned transition in 2012 is, in fact, all around us. This leap is actually not quite as big as one might think. The ancient knowledge of the Maya has been with us all along, stored deep down in our subconscious minds. To bring this knowledge back to the surface in order to prepare us for major changes that will come our way, we are merely being reminded of it now by other sources. These sources are all around us to guide us toward the evolution of 2012. Their information will manifest itself to everyone when we pay attention to what is happening in both our visible and invisible reality. This information is accessible to everyone who opens up to the energies that come from another dimension. Now, let me tell you how I came to know some beings from another dimension: the Arcturians.

THE ARCTURIANS ON 2012
Several days after the Harmonic Concordance of 2003 (when a lunar eclipse occurred in which five planets formed a Star

of David), I experienced a visit from the Arcturians. The Arcturians are multidimensional beings who live in a dimension parallel to our reality. They are benevolent ultraterrestrial beings who can converse with our intuitive senses, rather than being physically seen or heard. During their visit, I saw many beautiful colors in geometric shapes.

Inspired by the Arcturians' messages, I began developing art that resembled the images I received in my mind's eye. After posting the art on my Web site, I learned from those who saw it that, unknowingly, I had created art that replicated crop circles.

Like the Maya, the Arcturians have inherent, ageless messages to pass on to us. One of the main ways they communicate with us is through crop circles. This is also how they pass on the keys to help us toward a return of consciousness that is coming in 2012.

The Arcturians are responsible for the appearance of the majority of crop circles all over the world. As shown indisputably by well-known researchers Freddy Silva, Robert Boerman, and the BLT Research Team, Inc., many of the more complicated crop circles could not possibly be human-made. Scientist Paul Vigay, in particular, has encountered many aspects of strange and anomalous phenomena. He was one of the discoverers of unexplainable electromagnetic effects in and around crop circles.

The Arcturian circle makers leave information for us that will enhance the growth of our consciousness leading up to the transition in 2012. Their wisdom is encoded in these mystical, yet perfect, geometric shapes and forms. Every year, magnificent circle formations appear around the world in places where Ley lines cross. Ley lines, or Leys, are alignments of ancient sites across the Earth. These areas breathe extremely powerful energy.

The geometry found in crop circles is often called "sacred geometry," otherwise known as the ageless patterns and shapes that appear naturally throughout the universe. Ancient traditions like those of the Maya, Egyptians, and Native Americans, among others, recognized these forms and even discovered the numerical and supernatural significance of sacred geometry. These civilizations modeled temples and structures to resemble these sacred geometric patterns. Sacred geometry is innately part of the cosmos; all living creatures consist of geometrically shaped elements, even down to the cellular level. In this sense, sacred geometry is the most universal language that Earth has ever known. The geometric patterns within crop circles resonate within our subconscious, and our subconscious recognizes the messages, even if our five senses do not.

The purpose of the Arcturians' "supernatural" communication through crop circles is to guide humankind in its evolution to a higher level of consciousness. By using the energy, ancient symbols, codes, and holographic blueprints apparent in the crop circles, the Arcturians help prepare us both consciously and subconsciously for new times ahead, and especially for 2012.

THE HIDDEN CODES IN CROP CIRCLES

Over the past few years, crop circles have become much more subtle and complex. The reason for this is the expansion and growth of humankind's consciousness. As 2012 approaches, humans are slowly but surely opening their minds to the knowledge that is stored within these formations. While watching these sacred geometric forms and tuning into the energy present within the formations, our consciousness grows. The more complex the circles become, the more our consciousness is stimulated to grow. The information within the circles finds a way to reach our subconscious minds, even if we are not physically aware of it. If we remain open to this, we will develop our higher purposes in a variety of ways: physically, mentally, emotionally, and spiritually.

In a truly wonderful way, we are being guided to communicate with other forms of life that exist parallel to our own. The shapes and figures of the formations contain information about their origin. We are not yet capable of decoding all of this information with our brain or analytical thinking methods. As humans, we are quite visually focused. This can be a disadvantage, since comprehending the information coming from the crop circles is more easily understood and absorbed subconsciously, rather than by relying solely on our eyes and ears. Simply by focusing on the center of a formation, the information hidden in its code will be stored in our subconscious minds. They will help us to release pain, pressure, doubt, fear, and grief. The energy within resonates with our own geometric blueprint and seeks to heal pain we may have or may face. It is important to realize that crop circles are here to reflect the extraordinary beauty and wisdom present in every human being. They will help us and guide us in fully understanding all aspects of creation.

DECODING THE MESSAGES

I help others experience the extraordinary information in the Arcturian messages through my art. My art emulates the shapes of these crop circles. I give visual presentations around the world and show my art to an open, willing audience. Amazingly, many people have found the messages to have a transformative effect on their lives, and this will allow them to prepare for the coming changes in 2012.

Through these designs, I pass on what these crop circles signify with their sacred geometry. As seen in Figures 26

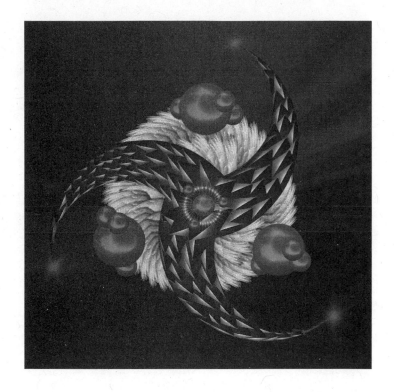

FIGURE 26: "Surrender."

and 27, surrender and empowerment are the messages the
Arcturians have passed on to me. I have translated 45 keys, or
messages, like these. They can be seen through my art on my
Web site, www.the-arcturians.com, and some may be seen in
my book and kit, *Sacred Geometry: Unlock Your Potential with
the Keys of the Arcturians.*

EXPANSION OF CONSCIOUSNESS
December 21, 2012, is fast approaching. According to the
Arcturians, on this date the transition to the fifth dimension

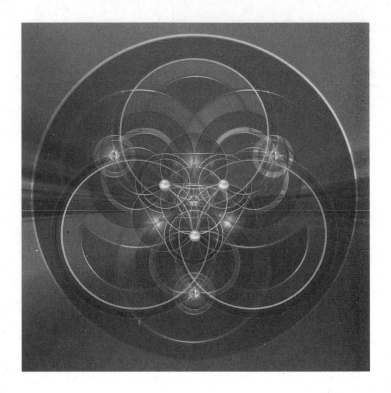

FIGURE 27: "Empowerment."

of our evolution will begin: the transition to a new Earth, a new harmony.

Time is accelerating toward 2012. As this happens, we are learning to understand what creation means and what our own active part in it is. We are all capable of creating our own reality, our own path in life. We are contributing to creating this reality by our words, thoughts, and actions—all of which carry their own energy. Because we are capable of creating our own reality, we should be aware of where our thoughts are taking us. For starters, in order to create and manifest from the heart, we should clean up the old habits and patterns that keep holding us back. Only then will there be room for pure creation.

You must ask yourself if you will take ownership of your responsibility to create your own future. This is a time for action, when all humans are partly responsible for creating the transition into our new reality in 2012. We are doing so by our own words and thoughts. Are we choosing drama, or are we choosing challenges? That is entirely up to us.

SACRED GEOMETRY AND THE HOLY GRAIL

Humankind and mother Earth are on the eve of a massive evolution. In 2012, the Age of Aquarius will commence; the Age of Pisces will pass. By expanding our consciousness and becoming more aware, we will view reality in a different way in years to come. You may have noticed that you feel trapped, stressed, and unhappy more often than before. Influences such as our past, work, and relationships slow down our development and cause us emotional and spiritual blockage. More often we think about our purpose, our goal on Earth: why are we here, and where are we going?

Around 2012, many will find their answers. The power of sacred geometry will help us understand what is going on and guide us subconsciously toward the future; a future where we will become one. This oneness is reflected in the strong connection between the power of geometry on our subconscious minds and the quest for the holy grail. When I refer to the holy grail, I refer to our quest for the ultimate mystery behind our existence, the remembrance of who we are and where we come from. This is what we all desire to know.

Sacred geometry can lead us to our own personal holy grail, if only we open our trust. Sacred geometry is not only to be found in crop circles, but in everything else around us and even inside of us. From the alignment of the stars and ancient architectural structures to every cell in our body and even our heartbeat, geometry is everywhere. The universe, the Earth, and everything living are created according to the golden ratio, the ultimate geometric proportion of divine perfection. By surrendering our stubborn reluctance, admitting that some things cannot be seen, heard, or touched, and that ultimate truth or higher consciousness only occurs when we let go of material reality, we entrust ourselves to the power of sacred geometry. Crop circles are not the only things that carry this meaning. Everything in the natural universe is singing this message of truth. If you look closely, you will see it, in trees, flowers, and your own body.

This universal geometry resonates with our subconscious and has a healing effect on everyone who opens up to this knowledge that is already stored inside us. It is the blueprint of our creation. It makes us who we are, our true selves. This is what connects us to the search for the holy grail. The vital magic of this legend is in fact the ultimate quest for the mystery of who we are. We don't view things as *they* are, we view things as *we* are. We are indeed powerful beyond measure. It is this power, combined with our own inner wisdom, that is capable of flowing smoothly into the new era of 2012 that is soon to come.

A Time to Remember

CHRISTINE PAGE, M.D.

It is said there will be a return of the feminine in 2012, portrayed in the symbols of the virgin, the mother, and the crone. This emergence of the feminine inherently implies a death, or diminishing, of masculine energy. According to author and teacher Dr. Christine Page, this change of tides, or shift in feminine/masculine power, can be seen as "the completion of the hero's journey." In the following essay, she illustrates not only how the myths of the feminine and the masculine incarnation relate to our personal journey toward 2012 and a new consciousness, but also how we can use the lessons of these symbols and myths to prepare ourselves for the coming changes.

Congratulations—you are among a select group of souls who won the lottery to be here, on this planet, at this time! The prize not only ensures you a front-row seat but also the unique opportunity to cocreate the future of the human race. Your contribution, along with that of other awakened souls, will create a blueprint, which will influence your ancestors and the next 26,000 years of human existence. This is what you have been working toward during your many incarnations; this is the moment you've been awaiting. This is a time to remember.

For the first time in 26,000 years, our sun and, by implication, the Earth are aligned with the dark rift of our galaxy, the Milky Way. This path of interstellar dust is known as the "black road" by the Maya and leads us directly to the galactic center.

According to the ancient people who observed the skies night after the night, the Milky Way represents the great mother and the great serpent, both synonymous with death and rebirth, the essential ingredients of transformation. The dark rift is seen as a symbol of both the open vagina of the mother and the gaping mouth of the serpent. Whatever the symbol, the message is the same: This is the portal through which we are born, and it is to here that we will return to die.

There has rarely been a more important time to understand this truth as we complete the last phase of what the Maya know as the world of their fourth sun and as we await the birth of a whole new world at sunrise on December 21, 2012. Metaphorically, we are now being inextricably drawn farther and farther along the dark road under the control of the black hole located within the galactic center. We are in the last death throes of the old era, and the hologram that we call reality is dissolving, causing many to experience *the anxiety of the unknown*, often translated as "chaos." At the same time, all of our creations from the past 26,000 years are returning to us so that we can garner the golden nuggets of wisdom and release the burden of the surrounding stories. Through this process, our inner light becomes stronger, easing our passage along the dark rift until we enter the galactic center and the mystery contained within.

To the indigenous people, this center represents the heart of the great mother. It was here that king-priests and shamans would travel to receive the insights and wisdom that could be translated into reality for the well-being of the land and its people. It is seen as the portal to multidimensional reality, parallel worlds, the void, the place of nothingness, and the ocean of possibilities, accessible in the past only to the elite or those in deep meditative states. Now, at this extraordinary time, its riches are available to all of us if we can step outside our limiting belief patterns and rekindle the innocence and joy that allow us to harmonize with the pulse of the great mother.

We are like children in a candy store, and the great mother is offering us an unlimited array of goodies. Yet, if our hearts and minds are still locked into a particular paradigm of belief concerning ourselves and the world, all we will see and select are the candies with which we are familiar, and we will pass by the rest. This is certainly a good moment to take a risk.

THE IN-BETWEEN TIME

The great shift in consciousness began in the late 1980s when, astronomically, the sun began its 36-year passage across the mouth of the black road. This time period encompassed both the Harmonic Convergence of 1987 and the close passage of Neptune and Uranus through the constellation of Capricorn in 1989. This latter event brought about the fall of the Berlin Wall, the collapse of the USSR, and the dissolving of other restrictive regulations, bringing freedom to many indigenous people and giving them the right to practice their own sacred rituals.

The window of opportunity to experience the abundance and ocean of possibilities accessible through the galactic center will be available until approximately 2023, when the sun will finally turn away from the black hole and start the next 26,000-year cycle of self-discovery and creative opportunity.

Metaphorically, this 36-year period could be symbolized by the monthly disappearance of the moon from the night sky. During this time, the old moon dies, making way for its son and heir, who will eventually appear as a thin silvery crescent three days later. In the same way that we cannot see the actual death and rebirth of the moon, so the transition between the worlds may not be outwardly defined. Yet to the Maya, at sunrise on December 21, 2012, the world of the fifth sun will be born, fully revealing itself around 2023.

WHAT DOES THE FUTURE HOLD?

The prophecies say that we are entering an era governed by the element of ether. Often described as the synthesis of the four material elements—earth, fire, air, and water—ether is celestial, lacking material substance, and existing beyond the confines of time and space. Yet, its nebulous nature should not be underestimated, for it is no less real than wood, wind, flame, stone, or flesh. Within the context of ether, there is a fusion of polarities that eliminates the need for separation into darkness and light, negative and positive, or good and bad. All are accepted as merely another expression of the same essence, lovingly created in the presence of the great mother. Such a concept certainly challenges those who prefer to believe that a unified field means that everybody will be the same, using prototypical ideals based on color, culture, religion, or gender. In the new world, the unification offered by the element of ether requires the acknowledgment of unity through diversity, where *every aspect* of the creative source is respected and honored without judgment, bias, or isolation.

What is now becoming clear is that such levels of acceptance will not be achieved through preaching, prayers, or patronization, but through the extraordinary waves of compassion that are already beginning to engulf our planet. Such divine compassion is unemotional, not attached to its acceptance, and is available to all, whatever their habits, thoughts, or beliefs. Yet, we can only embrace its presence when we step beyond a need to be self-deprecating and the belief that we are unworthy to receive. Only then will we know that love has no end.

This shift in perception is the first sign that we are truly facing the dawning of the Age of Aquarius, symbolized by enhanced social conscience, community cooperation, self-responsibility, and the ability to see the "bigger picture." It is transmuting egocentric behavior, guru worship, loss of personal power, and nonaccountability for one's actions, all of which are common features of the dying Piscean Age.

Once we can step beyond our self-obsession with emotions such as guilt, blame, judgment, and victim consciousness, we are given the opportunity to experience the true meaning of compassion. This is because, within the medium of ether where time and space are collapsed into *the now*, all our actions are simultaneously met with a reverberating response so that we automatically experience the effect of our deeds upon any person they impact. Hence, in the fifth world, we will find there is no separation between ourselves and the focus of our attention. Our thoughts will produce an effect to which we are instantaneously sensitive. In essence, we will no longer be able to rest in the belief that our thoughts and emotions are safely hidden as long as they continue to be unexpressed and unspoken.

As our commitment to the collective consciousness strengthens, we will come to truly appreciate the phrase, "Do unto others as you would do unto yourself," for now we will understand that there is no differentiation between the self and others. In other words, whatever I do to another, I am simultaneously doing to myself, *for you and I are one!*

From this place of understanding, it is very clear that wars would end, abuse would cease, and shameful secrets would be a thing of the past. There would be no purpose in hurting others when we would be merely harming ourselves. As a wise teacher once explained to me, "Without the separation maintained through fear and judgment, we will truly come to know compassion. Then we will experience spiritual telepathy, where all minds and hearts meet, united by a common cause."

WHY ARE WE HERE?

Now that we have previewed the future, it would probably benefit us to understand what we've being doing here on Earth for the past 26,000 years. According to alchemical teachings, the science of the mystics, the human experience is one of continual cycles of death and rebirth, where Earth rises into heaven and heaven descends onto Earth.

> *As Below so Above, As Above so Below*
>
> —*Emerald Tablet*

This is not a one-way journey; rather, it consists of cycles and spirals passing in both directions where *up* is no more important than *down*. The ability to pass with ease between these dimensions will bestow the traveler with the gift of immortality, a talent commonly believed to be reserved only for the gods and goddesses. Yet, all the great masters of the past have come to share the same message. They appeared on Earth to remind us that, first and foremost, we are divine beings, whether within our ethereal or our material states. From their own experiences, they appreciate how easy it is for us to lose our way, handing over our power to authority figures who are only too happy to see themselves as gods. At this important moment in Earth's evolution, the ascended masters are calling on all of us to wake up and remember who we are.

Over the past 26,000 years, we have moved many times between our spiritual ethereal selves and our dense material selves in order to achieve self-realization—in essence, to know ourselves as gods. The good news is that we don't have to wait for this life to end to experience these two worlds. We pass along the ladder between these different dimensions every time we enter the dream state, whether this occurs during sleep, meditation, daydreaming, the moon time, or between breaths.

During each of these events, our old selves undergo a mini-death, opening our hearts and minds to the new inspiration. Despite the reluctance of some individuals to accept physical death, the process of death and rebirth is natural and could be said to be as easy as breathing. A useful analogy is to imagine that our creative spark, or mind, is a fishing line. Every time we leave the known and enter the dream state, we are given the opportunity to cast our line into the ocean of possibilities, also known as the collective consciousness or imagination. These vast arenas of knowledge contain unrealized potential similar to the words in this book waiting to be brought to life by the creative spark of the reader.

As we reel in the new ideas and examine our catch, we feel an urge to transform these insights into manifestation; hence, the *hero* is born. This is the masculine face of the explorer, found within all of us, who steps out into the world, nurturing his ideas until they reach fruition. As a well-known archetype in modern mythology, his adventures and challenges are at the core of many great folk stories, inspiring others with his daring deeds. Eventually, the hero's trials end when he is crowned *king*, celebrating his achievements and standing proudly in the center of his creations.

Despite what many believe to be success, the fulfillment of a soul's destiny requires more than external achievements. Its energy must be transformed internally until it is available in a mode that can be used to enrich the collective consciousness.

For this shift to happen, the archetype of the king must give way to that of the *lover*, who turns away from the bright lights, toward that which will bring expansion to the light or Ka body, the vehicle of the soul between the worlds. Without a strong Ka body, we are destined to throw our fishing line into the same stretch of water over and over, believing that life just happens to us and that we are deprived of choice. As our inner light becomes stronger, we are offered the chance to not only move with ease between the worlds, but also to cast our fishing line farther and deeper into the unlimited ocean of possibilities.

So with sensitivity, compassion, and dexterity, the lover's self-reflection slowly dismantles the carefully constructed stories, responsive to the amount of effort expended in their creation, until only the essence is left. Here, faced with the raw core of the story, where all masks and veils have been removed, the lover embraces the pure essence without judgment and with deep compassion, eventually releasing the light of wisdom that feeds the soul.

Such moments of clarity and release often appear with the exclamation "Ah ha!" or with statements such as, "The light dawned; everything became clear," or "I got it!" The lover has now earned the distinction of becoming the *sage*, an ethereal counterpart to the king, and with the sage's appearance we experience:

- Release of the old story
- Lightness of heart and mind
- Completion of old patterns of attraction so that certain types of people and situations seem to disappear from our lives
- Choice as to where to cast the fishing line next time
- Greater ease in seeing the bigger picture

- Return to the great mother's ocean of possibilities, the void, and the timeless place of nothingness.

COMPLETION OF THE HERO'S JOURNEY

At this monumental time in our history, there are few if any individuals who are on this planet merely to begin a whole new journey. Most of us are here to finish what we committed to 26,000 years ago. There are certain guidelines to check to see where the journey may have been curtailed or where we have become trapped in a cul-de-sac of delusion and need to extricate ourselves to complete the cycle and bring the essential light to our Ka body.

The Hero Phase
- What dreams and ideas do you hold in your imagination that you've never had time to nurture?
- What regrets would you have if the world were to end tomorrow?
- What seeds of desire were planted and then deserted for some other project?
- Is it time to take a risk in your life?

The King Phase
- How easy is it for you to celebrate your achievements?
- How comfortable are you wearing the crown of success?
- Of what three achievements are you particularly proud?
- How would your relationships change if you owned your crown?

The Lover Phase
- How much of your identity is bound up in external appearances and commitments?

- Who are you when you remove the veils and the masks?
- How easy is it for you to offer compassion to all aspects of the self, including those that bring you shame or embarrassment?
- Can you love the unlovable parts of yourself?

The Sage Phase
- What patterns of behavior are dead and need to be released?
- What relationships would not survive your transformation?
- What beliefs about yourself are holding you back from experiencing the fullness of life?
- How easy is it for you to release control and surrender into the arms of the great mother?

As we complete each cycle of self-discovery, we extract the pure essence of light and add this to our own Ka or light bodies. Over time, this subtle body develops its own magnetic force and starts to attract, at an increasing speed, people and events that will help it on its journey, which we call synchronicity. Our eventual goal is to be able to live on Earth entirely through our light bodies, no longer dependent on our physical ones for survival. In this way, we would know the immortality that, until now, was only the domain of the few, including masters such as Jesus, El Morya, the Virgin Mary, St. Germain, Buddha, Muhammad, Djwhal khul, and Kuan Yin.

THE RETURN OF THE FEMININE
Over the past 2,000 years, much attention has been given to the masculine face of divinity, although the truth is that the transformational process can only take place with the full cooperation of the divine feminine within each of us. She

alone can provide the alchemical vessel that embraces the mystery of death and rebirth, qualities that the masculine aspect often fears and attempts to avoid at all costs. Even in apparently liberated countries, her full face is still commonly hidden, although this was not always the case. In most ancient traditions, the great goddess was revered, with even the gods bowing their heads in her presence.

There are many mythological stories that exemplify her importance, although the following tale comes from relatively modern times. During the creation of the Iroquois Constitution, which became the basis for the Constitution of the United States, it was decided that the chiefs would be chosen by the women of the tribe. This reasoning emerged from the profound understanding that only women knew the hearts of men, only women were connected to the Earth's abundance, only women knew how it felt to bury their loved ones, and only women could decide whether any battle or war was worth the cost of life. Imagine the global reformation that would occur in both politics and religion if *only* women had the vote.

The feminine face of the divine is commonly described by ancient cultures as the triple goddess, composed of the virgin, mother, and crone. Each is seen to have a particular role in the creative process:

The Virgin—Creator

The Mother—Nurturer

The Crone—Destroyer

Examples of the three-faced goddess include Parvati/Durga/Uma (Kali) from the Hindu tradition, Ana/Babd/Macha (the Morrigan) in Ireland, and Hebe/Hera/Hecate in Greece. She also appears as the triple aspects of Guinevere in Arthurian legend, as Diana Triformis to the Druids, and as the Fates among the Romans.

In order to energize each aspect of the feminine, it is important to speak of them separately, although in essence they work concomitantly.

THE VIRGIN

The precise definition of this archetypical energy is "to be complete unto oneself." Such a state of completeness makes it possible to understand why in mythology the virgin is often reported to be unmarried, for she "has no need of another to make her whole." Her clothes of white emphasize her ability to radiate authenticity; nothing remains in the shadows, and all aspects of the self are embraced in the name of love. In essence, anybody who exemplifies the virgin is a definite threat to those who gain control of others through manipulation, shame, guilt, and fear, for she cannot be coerced through emotional blackmail. This is the reason why, over the past 2,000 years, her image has so often been corrupted so that she is portrayed as immature and untainted by the joys of life.

Yet, the impact of the virgin on our existence can never be erased. She is our higher self carrying the spiritual blueprint that we have chosen to bring into manifestation during this life. Her greatest desire is that we come to love and accept all aspects of ourselves so that we, too, can radiate the pure light of authenticity. With this in mind, she speaks to us through the small voice of intuition, helping us navigate our way through the eddies of delusion and around the rocks of stagnation. She provides simple guidelines to help us maintain our focus on fulfilling our soul's destiny. These include:

- Am I acting through love or fear?
- Does my job/occupation nurture my soul?
- Do my relationships allow my soul to grow?
- Am I excited by my life?

The virgin never deserts us, staying with us during our darkest moments, when doubts or fears nearly defeat us or when crossroads appear in our life. She knows our blueprint intimately and gently nudges us forward toward our soul's destiny. This is more profound than we can imagine, for the virgin has a much deeper role, identified within the mythological tales of the Roman vestal virgins and the priestesses of the Virgin Brigid in Ireland. These young women were specifically trained to be guardians of the hearth and, in particular, to guard the *perpetual fire*, which was never to be allowed to be extinguished.

And what was this fire? The Sufi master Hazrat Inayat Khan offers the clearest understanding to this question in *The Spiritual Message of Hazrat Inayat Khan, Vol. 1, The Way of Illumination*:

> *If love is pure [and] if the spark of love has begun to glow, then there is no need to go somewhere to gain spirituality, [for] spirituality is within. One must keep blowing the spark till it turns into a perpetual fire. The fire-worshippers of old did not worship a fire which went out, they worshipped a perpetual fire. Where is that perpetual fire to be found? In one's own heart.*

The Romans and other ancient people worshipped the heart, whether of an individual or a civilization, knowing that without it, growth and creativity would quickly become stagnant, leading to the eventual death of the organism. They knew that the heart was more than a simple pump, and they saw it as a transformer of energy, transporting an individual or a culture through the various realms of multidimensionality until one could choose to live outside the confines of time and space and know immortality. Most of us have experienced such freedom in our lives during moments of joy, heartfelt happiness, and within the bliss state of early love. Time seems to stand still, our sensations are heightened, colors become brighter, and we feel as if the abundance of the world is being offered to us on a plate.

It is this same state of awareness that we are presently facing as we align with the heart of the great mother at the center of the galaxy. All that is required is for us to provide her with enough love to fuel the fire that will act as the impetus for our travel into the timeless void, with its ocean of possibilities. And this is where the virgin enters. Her greatest interest is for us to exist in our completeness—where there is fusion between our inner and outer lives and between our worlds of spirit and matter. She knows that as we manifest each aspect of our spiritual blueprint into reality and then draw its essence into our hearts in the name of love, our fires become brighter, our light bodies stronger, and we can move with ease between the dimensions. The burdens that prevent such movement include the baggage we carry from the past, unfinished stories, delusions of the future, and attachment to the belief that we control our lives. It is through the virgin's guidance that we eventually release our hold on our redundant reality and step with courage into the unknown to once again be warmed by our own perpetual fire.

THE MOTHER

Of all the archetypes, the mother image has been the least corrupted over time, with most people appreciating the gifts of nurturing and the abundance she offers every moment of our lives. She is the ocean of possibilities into which we cast our fishing line, the nourishment that feeds the hero on his journey, the throne on which the king sits, the cave that the lover must enter to know himself, and the waters that welcome the sage home at the end of his travels.

Some mythological tales tell of her grief when her child is lost or leaves home, yet the great mother's love is so deep that in such grief there is also celebration as her child sets out to fulfill his or her destiny. Is this not true of any parent? Isn't it the power of love that causes the mother bird to push its offspring from the nest and trust the inherent quality of flight?

The mother's abundance is always available to us as long as we approach her with love, honor, and respect—qualities that feed her fire. When we fail to appreciate her needs, the doors to her gifts will close, and we will be left to feel the pain of her departure. It is vitally important on our planet at this time that we learn to respect and honor the mother, whether we see her in the food we eat, the energy we use, the furniture we sit on, or the physical bodies that carry us through this life. It must be apparent to everybody that she is slowly withdrawing her assistance as this old world dies and before the new one is born. However, the abundance of the new world is based on our ability to appreciate the great mother *now*, and therefore this is a great opportunity to

- Celebrate her offerings with reverence.
- See sacred sites or places of sacredness as her home, and visit bearing gifts of joy, laughter, song, and dance.
- See the trees and plants as her lungs, without which we will fail to be inspired.
- See the rivers and oceans as her blood, which must be kept clean and unpolluted if they are to continue to nourish us.
- See the rocks and mountains as her bones, which offer us security and stability.
- See as our family all the other creatures on this planet, each of whom reflects part of our human nature.
- See each human as another part of ourselves, and meet him or her with love, honor, and respect.

THE CRONE

The final character in the trinity is probably the hardest to integrate, and yet the most important at this time of dissolution and rebirth. The crone, or dark goddess, is present in all cultures and is often depicted as a witch or an ugly old woman surrounded by dark animals and in charge of a large fiery cauldron. She appears as Kali the destroyer, Cerridwen the corpse-eating sow, Sekhmet the fire-breathing lion, Isis the vulture queen, Morgan Le Fay the death queen, and Persephone the destroyer. It is no wonder she has acquired a reputation of death, representing winter, disaster, the waning moon, and doomsday, and always reminding us that regeneration only occurs through dissolution of the old.

Now, at the close of the 26,000-year cycle, she is the black hole in the center of the galaxy, which is drawing us ever closer, and her vagina is represented by the dark rift. We need to remember that it was our higher self that called on her to help with the transformation process, and there is no going back.

As the king accepts that his destiny is to die so that eventually his son and heir can be born, the lover enters the underground womb or cave of the dark goddess. This shift in consciousness is commonly described as the dark night of the soul when, contrary to everything that is happening in the outer world, we inexplicably turn our focus within and seek the space to be introspective. Understandably, family and friends often express surprise at such a move, unaccustomed as we are in Western cultures to give time to this essential part of the creative cycle.

Seen as an episode of depression, "happy pills" are prescribed to lift us out of the darkness, and encouragement is given to motivate us back to work. Yet, all ancient cultures saw the inner work of the descent into the underworld as essential for the survival not only of the individual but also of the tribe. All were encouraged to visit the dark goddess through practices such as vision quests, walkabouts, and, for women, meditations during their moon time.

In the darkness of her cavern and with her cauldron of fire, the goddess assists the lover in dismantling the stories that have been created around an experience so that the true reason for the creation can be discovered and the essence of wisdom released. This part of the journey is not easy, because we have often become so attached to the worlds we have created and the stories we tell ourselves. We see the crone's methods as crude and unloving as she eats away at the flesh and entrails of our stories like a vulture and boils us in the fires of her cauldron. Yet, her only purpose is to reach the dark places of our psyches, where shame and fear prevent us from loving ourselves fully and, with the aid of our inner lover, fully embrace those disowned parts into our hearts. As each abandoned aspect of the self is returned to the heart, our inner light increases and our Ka bodies grow stronger.

Despite our higher awareness, our smaller self will use every trick it knows to hold on to the last vestiges of the dying world, failing to appreciate the tremendous love the dark goddess offers in her desire to assist us to know ourselves in our true light. What we see as death, she sees as rebirth into the light.

Finally, however, the goddess's love prevails, and we find it in our hearts to love even the crone, for she is a part of us, and we feel the peace that emerges as the sage surrenders into the warm waters of the great mother. In time, he will choose to once again cast his line into the great mother's vast ocean of possibilities and create anew.

THE WAY AHEAD

We are in one of the most creative periods of the Earth's history, when the future is open to all manner of possibilities. Our responsibility is to be willing to love, accept, and integrate every aspect of ourselves into our hearts, our perpetual fire, so as to generate the energy and light required to travel into the void, with its multidimensional possibilities. The heart of the great mother awaits us; our intuition will guide us, and our lover waits within. All we need to do is remember why we are here.

2012, Galactic Alignment, and the Great Goddess

REFLECTIONS ON ISIS AND THE SACRED SCIENCE OF THE EGYPTIANS

SHARRON ROSE

Galactic alignment is at the center of astronomical predictions for 2012. Along with a shift in consciousness, it is widely believed that 2012 and the galactic alignment herald the emergence of the goddess energy. Sharron Rose traces symbols of the goddess Isis and Egyptian hieroglyphs as they relate to the phenomenon she foresees for the coming 2012 era. From the Black Madonna to various other incarnations of Isis, Rose reveals ancient predictions through symbols and myths surrounding the galactic alignment, its relationship to the Kabbalistic "Path of Return," and the sacred marriage (the reunion of female and male energies). She addresses the questions, How did ancient Egyptians interpret the galactic alignment? What did their interpretations mean through their sacred science?

Many have written on the symbolic relationship between the end-date of the Mayan calendar and the galactic alignment. I would like to focus on the galactic alignment of 2012 from a very different perspective, that of the symbolic and metaphorical teaderings of ancient Egypt, which formed the foundation and guiding force underlying all of Western mysticism.

Symbols, myths, and metaphors contain within them the potential to arouse numerous intuitive connections and associations within all of us. Throughout antiquity, symbols were utilized by traditional civilizations across the planet as part of their esoteric and initiatory teachings. Symbols were developed to open the heart and intuitive faculty, to evoke awareness of, contemplation upon, and insight into the forces that weave the outer fabric of the universe and the inner fabric of the human being. They were part and parcel of the body of knowledge that is known to us as sacred science.

By weaving together myth, symbol, science, and contemplation, it is my hope to bring you to a new awareness of the significance of the transformative power of the galactic alignment from both a personal and a universal perspective.

On December 21, 2012, the Earth will line up with the Sun and the center of the galaxy. From a symbolic perspective, what does this mean? In Part One of this essay, we will begin to unveil some of the esoteric teachings found at the heart of this galactic alignment by exploring the ancient Egyptian myths and symbols associated with Isis and the center of the galaxy.

Anedj   hr (hair)   Auset   uret   Hekau-t.
Hail to thee Isis, Great One of Magic (or Words of Power).
FIGURE 28

PART ONE: ISIS UNVEILED

Once upon a time, in the long lost days of Al Khem, or ancient Egypt, there was a temple at Sais dedicated to Isis. Inscribed on the altar of this temple were the following words:

> *I am she who separated the heaven from the Earth. I have instructed mankind in the mysteries. I have pointed out their paths to the stars. I have ordered the course of the Sun and the Moon. I am queen of the rivers and winds and sea. I have brought together men and women. I gave mankind their laws, and ordained what no one can alter. I have made justice more powerful than silver and gold. I have caused truth to be considered beautiful. I am she who is called the goddess of women.*
>
> *I, Isis, am all that has been, that is, or shall be; no mortal man hath ever me unveiled. The fruit which I have brought forth is the Sun.*
> —Inscription from Isis Temple at Sais

For the Egyptians, the center of the galaxy was symbolized by the great goddess Isis in her role as the mother of creation. She was the *prima materia* and the secret womb of all that is. She represented the source of the great outpouring of cosmic rays, dust, and other rare elements that make up the material that our scientists have detected spewing out from the center of the galaxy.

The huge gravitational pull of this center produces a vorticular-shaped galactic formation that looks to us on Earth like a pinwheel. This vortex of stars is continuously emerging from and spinning around the center. The vast curving arms of the Milky Way are formed from the dust and particles that emerged from the center of the galaxy. The center of the galaxy is so immense that its size is more than one million times larger than our Sun.

One of the great secrets of ancient Egypt is that the center of the galaxy is the secret sun that exists behind the Sun. In alchemical language, the center of the galaxy was also called the black sun, the hidden sun, or the invisible sun. It was the secret sun of the eternal feminine, the *soror mystica*.

For the Egyptians, Isis was the creator of all things. They wholeheartedly believed her declaration, inscribed at the Temple of Sais more than 3,000 years ago, which also said,

> *I, Isis, am all that has been, that is, or shall be.*
> *The fruit that I have brought forth is the Sun.*

Viewed through the lens of symbolic thinking, a statement such as this and the image of Isis herself may contain infinite levels of information, hidden to our untutored eyes.

It is worth noting here that the word *sun* as written on the Temple of Isis at Sais is spelled s-u-n, referring to the great yellow star that nurtures and warms our planet. Egyptian myth tells us that Horus, the divine child seen suckling upon her breast, is the son of Isis and her beloved consort Osiris. But according to the ancient Egyptians, Horus also symbolizes the Sun that shines in our sky. Through this simple use of symbol and metaphor, the Egyptians are telling us that Isis, the center of our galaxy, gave birth to our radiant yellow Sun.

Horus, the son of Isis and Osiris, is both the "son" and the "Sun." Horus is the representative of his father, Osiris. The energy of Horus within us, awakened and invigorated by Isis, becomes the rising up of the will to complete the great work and attain full enlightenment in this lifetime, a process that the great Egyptian scholar Isha Schwaller de Lubicz called "willed superevolution."[1]

Centuries later, the myth of Isis, Osiris, and Horus was recast by the Christians as the Virgin Mary giving birth to the son who redeems humankind. Through the mystic story of the Virgin Mary and her son Jesus, the myth of Isis and Horus, of the great mother giving birth to the child of light, was reborn into our modern age. It is a strange and fascinating aspect of Christianity, this recasting of the mythic figures of Egypt into new symbolic expressions of the old gods and goddesses.

As the teachings of ancient Egypt, the birthplace of the alchemical tradition, spread to Europe, the myths and symbols of the great goddess Isis went with them in the form of the Black Madonna. In his masterful magnum opus, *Les Mysteres des Cathedrales*, the enigmatic French alchemist Fulcanelli stated, "Formerly the subterranean chambers of the temples served as abodes for the statues of Isis, which, at the time of the introduction of Christianity into Gaul, became those *black virgins*, which the people in our day surround with a quite special veneration." He goes on to tell his readers that the black virgins represent in hermetic symbolism "the *virgin Earth*, which the artist must choose as the *subject* of his Great Work."[2]

Frequently, the Black Madonna's were made of volcanic black rock that comes from deep in the Earth. From this perspective, the Black Madonna represents the prima materia, or first substance, the original substance from which all alchemical transmutation manifested. Thus, she arose directly from the very matter that makes up our planet. She holds her son in her arms, exoterically depicting the baby Jesus, but esoterically representing Horus, the son of Isis and Osiris. Hidden deep within the occult mysteries of the Christian tradition, we discover a symbolic representation of the center of the galaxy giving birth to the Sun. The fact that these alchemical myths are primarily astronomical in nature should not surprise us.

What does it mean when the myths say that Isis brought forth the fruit, which is the Sun? It is clear that the Egyptians may have known more about the true cosmology of our universe than possibly even we do today. To explain this, we need to take a look at the work of astrophysicist Paul LaViolette and his book *Beyond the Big Bang*. He studied data collected by NASA, the Hubbell space telescope, and hundreds of other sources to describe a new theory concerning the nature of how our Sun and galaxy interrelate. If his theory is correct, it indeed helps to explain the Egyptian symbol of Isis as the center of the galaxy.

As maintained by LaViolette, this area at the center of our huge spiral galaxy occasionally becomes active and literally explodes. This is a condition that astronomers call a Seyfert galaxy or quasar. According to this theory, there is a continuous transmission of gamma rays, electromagnetic pulses, cosmic dust, and other elements being thrust out of the center of the galaxy during its explosive stage. Literally everything that is of the solid galaxy, including all stars, planets, nebula, and comets, is the result of these diverse cosmic forces pushing outward from its center. In 1992, NASA's *Ulysses* spacecraft reported that most of the cosmic energy that it detected in outer space was coming directly from this source.

All galaxies go through this periodic explosive state. LaViolette calculates that the last time our galaxy became active was around 13,000 years ago. He believed that this burst from the center of the galaxy possibly lasted for 1,000 years. LaViolette tells us that from the Earth's perspective, this incredible emission of light would be shaped like a giant eye in the sky caught between Sagittarius and Scorpio.[3] Besides looking like an eye, to our ancient ancestors, this gigantic galactic burst also looked like a womb. This vision of a galactic womb brightening up the night skies for a thousand years or more may have given birth to the myth of Isis. In her appellation as the mother of creation, it is during this explosive phase that the face of Isis is finally revealed. After this brilliant birthing phase, after the explosive center has finally gone quiescent, Isis once again becomes the dark-veiled goddess whose face is once more hidden from our view.[4]

One of the most profound secrets of the ancient alchemists was the knowledge of the precession of the equinoxes. The zodiac was conceived to mark the great cycle of the Earth as it wound through its 26,000-year processional wobble. Oddly enough, the ice core drillings done by Dr. LaViolette in Greenland revealed that these periodic outbursts from the center of the galaxy occur in 26,000-year intervals. Perhaps due to some undiscovered mechanism, the rhythm of the wobble of the Earth matches the rhythm of the periodic galactic core outburst.

The Earth and our solar system are nearly 26,000 light-years from the center of the galaxy. We cannot see the center of the galaxy when it is in its quiescent phase. There are 26,000 light-years of dust, matter, and stars lying between the Earth—being out on the far edge of the spiral vortex of the galaxy—and the great center of our cosmic neighborhood. The center of our galaxy shines thousands of times brighter than any other part of the galactic body. Yet, due to the dust and debris, this wonderful presence lies hidden to our view. The dark veils of Isis are drawn between our planet and the center of creation.

As we have seen, the inscription on the temple of Isis at Sais states that the fruit she has brought forth is the Sun. Our radiant local star eventually emerged from the dust that escaped the throes of the fiery furnace, which lies at the center of all creation in our galactic neighborhood. Indeed, it can be said that everything that is, all of nature, human beings—even human consciousness—originally comes from this womb of the galaxy. If she is Isis, then our galaxy is the great mother, the creative, feminine force from which everything emerges, her pregnant womb continuously giving birth to millions of Horuses, or stars, that shine brightly on her heavenly body. This knowledge of the inner fires and compressed gases of creation, burned and transmuted in the crucible of her womb, lies at the heart of the Western tradition of alchemy. It is one of the fundamental secrets of this ancient and sacred Egyptian science.

Isis and the Virgin Mary are also associated with the Moon. As we shall see, there are interesting relationships between the cycle of the Moon and the great galactic cycles. Many European alchemical traditions used the annual 13-month lunar calendar. This calendar created a thirteenth sign, called Ophiuchus, or the serpent holder. Situated right at the galactic center between Scorpio and Sagittarius, Ophiuchus was the secret sign of the alchemists. The success of their alchemical experiments depended upon the quality of the moment in which the experiments were performed. Ophiuchus, the center of the galaxy, the secret sun, and the hidden sign of their occult zodiac, governed this quality of time.[5]

Ab-a   en   mut-a   Auset.
My heart, (is, of) my mother, Isis.
FIGURE 29

It is interesting to note that for early Christians, Mary Magdalene was the thirteenth apostle. From a symbolic perspective, she represents the secret Isis, the thirteenth sign, the center of the galaxy, a powerful creative force invisible to our view. This mysterious sign of the zodiac disappeared with the advent of the Roman solar calendar, containing only 12 months. The conquest of the solar tradition quickly replaced the ancient lunar/galactic tradition and ushered in the modern age of technology and machine.

We know that the galactic alignment involves the lining up of the Earth, the Sun, and the center of the galaxy. We have seen how a 13-month lunar calendar introduces a new sign into the zodiac called Ophiuchus, the serpent holder, that sits between Scorpio and Sagittarius at the heart of the center of the galaxy. But, how does the Moon relate to the galactic alignment and the symbolic teachings of the great goddess?

Dr. Wilhelm Reich speculated that the center of the galaxy was emitting waves of orgone energy that were strongest around the galaxy's ecliptic. Fortunately, our solar system lies in the ecliptic zone of the galaxy. This is the reason the Sun comes into alignment with the center of the galaxy.

The Earth and the Moon are, in actuality, one system. The Moon is not so much orbiting the Earth as the Earth and the Moon can be said to be dancing with each other. The Earth and the Moon spin around each other like a celestial barbell. This unusual interaction modulates the frequency of the electromagnetic field of the Earth.

The Earth is floating together with the Moon on the sea of waves emerging from the center of the galaxy. The Earth and the Moon are bobbing on this galactic sea, and this dynamic motion is called the procession of the equinoxes.

From an esoteric perspective, these galactic waves are emanations of the divine feminine streaming forth from the primordial heart of creation. This powerful feminine current, known in the tantric tradition as Shakti and in the kabbalistic tradition as the Shekhina, is the animating and nourishing force of life itself, the supreme all-illuminating power in every thing and every being. Like a spark of electricity igniting a film projector to send forth its beams of light, which then coalesce to create the flickering images on the cinema screen, this divine feminine force bursts outward from the supreme source in glittering waves of luminosity that merge and shape themselves into material reality.[6]

The 23.5-degree wobble of Earth's axis is the observational dynamic of the effects of these galactic waves. Like a buoy on the sea bouncing around during a storm, the Earth/Moon system bobs and weaves with the continuous flow of the light and energy of the divine feminine current emitted from the center of the galaxy. The pulse of the center of the galaxy, the heartbeat of the great mother Isis, modulates this wave of orgone energy. The resonant frequency of the wave comes in 26,000-year intervals. Every 13,000 years, on the solstice, the center of the galaxy and the Sun rise in alignment with the Earth.

This wave from the center of the galaxy modulates the spin of the Earth/Moon system. This central dynamic means that the Earth/Moon system is spinning in resonance with the waves being emitted by the center of the galaxy. The Earth and the Moon are like a pontoon floating on the galactic ocean, controlled by the powerful galactic waves and their dynamic pulsations. The Earth and the Moon pulse with the essential rhythm of the center of the galaxy or the heartbeat of Isis. This also means that the time signature of the lunar cycle is closely modulated by the larger galactic rhythm.

The sacred number that drives the Long Count calendar of the Maya is the 260-day tzolkin. As we have seen, the Earth/Moon system's frequency is tuned to the main harmonic frequency coming from the center of the galaxy. To the Egyptians, this was the heartbeat of Isis. It takes 26,000 years for one note in this harmonic to ring. The number 260 is a fractal of 26,000. This 26, 260, 26,000 fractal harmonic is the contributing factor to the emergence of human consciousness.

This harmonic fractal is also revealed by the procession coincidence. Not only does it take 26,000 years to go through one complete processional cycle, but the center of the galaxy is also approximately 26,000 light-years away from the Earth. Another astonishing coincidence is that the average human life contains approximately 26,000 days. From the perspective of sacred science, these coincidences are shadows or fingerprints of the larger galactic dynamics.

PART TWO: ISIS AND THE TREE OF LIFE
Reflections on the Galactic Alignment
and the Kabbalistic Path of Return
In Figure 30, a beautiful prayer from the Papyrus of Ani, the adepts are asking Isis, the universal mother and keeper of the sacred mysteries, to awaken within their hearts and open the star chamber. From a symbolic perspective, the star chamber, or seat of Isis, is the Milky Way and center of the galaxy. On another level, it is the pineal center and eye of gnosis (directly felt perception) of Taoist alchemy. In the Indo-Tibetan tantric tradition, it is the third eye, the wisdom eye, or Dakini eye. It is the secret inner eye that sees in dreams and visions, that sees beyond the veils of corporeal reality.

From a kabbalistic perspective, the star chamber is the hidden *sefirah, Da'at,* or knowledge that lies along the central pillar of the Tree of Life, or symbol of the world axis. Da'at is the receptacle for the combined Shakti/Shekhina current streaming forth from the first three sefirah—*Keter*, the supreme crown of God, pure and absolute essence; *Hokhmah*, or wisdom, intuitive instantaneous gnosis; and *Binah*, or understanding, revelatory

Auset, nehas em her ab un.
Isis, awaken within our hearts.

A-un akh-akh ar-et pafi (e)mtun ami ari
Kheperu ar Hru.
Open the star chamber that we may guide our
transformation into Horus.
FIGURE 30

and creative intelligence. Like a radiant crown of light sitting atop one's head, this trinity of sefirot channels the divine effulgence directly into the pineal center/star chamber. Whether in its appellation as the star chamber, Da'at, or the pineal center, it is the gateway to the union of Shakti and Shiva, the Shekhina and the Holy One, blessed be He, human and divine. When we enter the star chamber, the veil is lifted. We encounter the great goddess Isis in all her splendor, waiting to bring us face to face with divine reality.

Horus, who is both the son of Isis and Osiris and the Sun, the brilliant star that illuminates our world, also represents the point where the quickening within happens and the drive toward total

FIGURE 31: Akh—the spiritual light. The N (top glyph) denotes duality in motion. Adding this to Akh gives us Ankh, light in manifestation.

enlightenment in the current lifetime eclipses all other desires.[7] It is interesting to note that with the birth of Horus, the number of gods in the heliopolitan cosmology of ancient Egypt increased from 9 to 10, a sacred number that symbolizes completion and a return to the beginning. For it is through the opening of the star chamber of Isis, and the drive toward full integration, as represented by Horus, that the aspiration for and reunion with the one supreme source is realized.[8]

Isis is often pictured holding the Ankh, or the symbolic expression of the universal spiritual light as manifested in duality. As seen in Egyptian paintings and sculpture, it is through the Ankh that the vital, enlivening Shakti/Shekhina current is transmitted.

In Figure 32, we see how the Ankh is an early representation of the Tree of Life. The Tree of Life is the glyph, or composite symbol, that lies at the heart of the teachings of the Kabbalah.[9] Through personal contemplation on the symbols of Isis, the kabbalistic Tree of Life, and the Hindu chakra system, some fascinating insights into the nature of the galactic alignment have emerged. To fully comprehend these insights, one must understand the ancient Egyptian teaching of "as above, so

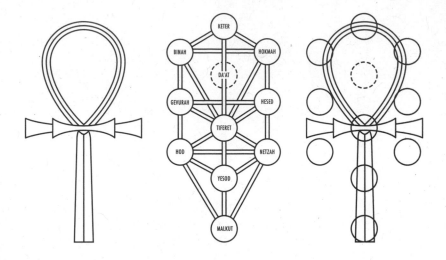

FIGURE 32: Ankh; Tree of Life; Ankh as Tree of Life.

below," or anthropocosmos: that which manifests in the outer world also manifests within the Earth itself and on the inner realms within our own beingness.

In the *Emerald Tablet*, the great Egyptian sage Hermes Trismegistos wrote, "Tis true without lying, certain and most true. That which is below is like that which is above, and that which is above is like that which is below to do the miracles of one only thing. And as all things have been and arose from one by the meditation of one: so all things have their birth from this one thing by adaptation."[10]

In every part and parcel of our reality, there is a great overarching pattern or structure that repeats itself from the macrocosm to the microcosm. In the words of the great Hebrew scholar and writer Rabbi Adin Steinsaltz, "Man may therefore be viewed as a symbol or a model of the divine essence, his entire outer and inner structure manifesting relationships and different aspects existing in that supreme essence."[11]

The Ankh, the Tree of Life, and the chakra system are all symbolic representations of this universal pattern or expression. Each has its unique characteristics, but what is most common to them and important in the context of the galactic alignment is the central pillar, the central channel or *djed* pillar, the direct path of emanation from and return to the divine source.

Looking at the Tree of Life, one finds at its base the sefirah *Malkhut*, or kingdom, the home of the Shekhina, the ultimate receptacle and full material reflection of the light and energy of the primordial source. Above her one encounters *Yesod*, the Moon, or the realm of emotion, sexuality, the lunar force and ebb and flow of the tides. Rising above Yesod is *Tiferet*, the Sun, the realm of beauty, the harmonizing principle and basis of the good, the compassionate, and the true. Above Tiferet is Da'at, the veiled sefirah, the abyss, the center of the galaxy and seat of gnosis. High above Da'at, like a crown of creation, is *Keter*, absolute transcendence, and the supreme unity beyond duality. When the galactic alignment occurs on December 21, 2012, the Earth (Malkhut), the Sun (Tiferet), and the center of the galaxy (Da'at)—indeed, the whole central pillar of the Tree of Life—will be in alignment. This portends a time of great opportunity for humanity.

According to the kabbalistic teachings of Isaac Luria, as the luminous radiance of the Shekhina poured forth from the divine source, the sefirot, or symbolic vessels of the Tree of Life, were unable to completely contain the fullness of this luminosity. Overflowing with this divine radiance, the first three sefirot—Keter, Hokhmah, and Binah—remained intact but still overflowed with light. The remaining seven sefirot were not strong enough to hold this pure light; the sefirotic vessels were shattered, and the light was broken up into sparks. These sparks fell through the metaphysical void into the *Sitra Achra*,

Uben – et ah   der   senekh.

I rise as light to repulse darkness.

FIGURE 33

the other side, interpenetrating the world of action and mate-rial reality into Malkhut, the world of the Shekhina, of earthly manifestation, of time and space. As moths are attracted to a flame, the sparks attracted layer upon layer of shells/husks known as the *kelipot*. The kelipot obscured the light and knowl-edge of our primordial source.[12]

This metaphor has a distinct parallel in the Indo-Tibetan concepts of the *kleshas*, or origins of human suffering. The kle-shas, emotional defilements or "poisons," bind us to the endless cycles of death and rebirth. Like storm clouds swirling across the sky obscuring the radiant light of the Sun, the kleshas of ignorance/delusion, anger/aggression, attachment/desire, pride/arrogance, jealousy/envy veil the innate clarity of the mind. To achieve true enlightenment, one must totally cleanse the mind of these defilements, transmuting them into the enlightened qualities of spaciousness, clarity, discernment, equanimity, and all-accomplishing wisdom. Through this powerful transforma-tive process, the veil is lifted, and one is brought face to face with one's true eternal nature.

In actuality, we are eternal beings. We are the reflec-tion of divinity, a fractal representation and mirror of the divine source. As seeds of the pure light and energy of the Shakti/Shekhina, we are the way in which divinity comes to know itself. We are god/goddess in the flesh, a materialized,

Nuk ua em ennu en Khu ammu Xu.
I am one of those of shining beings of light.

FIGURE 34

crystallized emanation and embodiment of our divine progenitor, the outer projection of the divine plenty (Shekhina) into physical reality (Malkhut). We are all unique facets of the complete image of an ever-evolving interconnected universe. Shrouded by the veils of karma, the husks of the kelipot, we have forgotten our divine inheritance. Yet, once the door is opened to this ancient knowledge, there is the potential for a great transformation to occur.

As maintained in the teachings and practices of all the great wisdom traditions throughout the planet, there comes a moment of awakening when the seeker turns her or his heart and mind away from the distractions and noise of the outer world of matter toward the inner world of spirit. Once this occurs, once one begins to recognize the beauty and grace of the larger divine reality, a longing for inner freedom, clarity, integration, and reunion arises. Through the process of purification, through devotion, commitment, and right action, the soul looks inward and begins to long for its home in the light.

According to the kabbalists, the most important task of every Jew is to participate in the *tikkun*, or the restoration of the world. In the *shevirah*, the shattering of the vessels or fall from grace, each spark of the Shekhina, each soul spark, experienced the pain and suffering of separation from the magical bliss of divine union. From one perspective, we fell to bring light to the darkness, enlighten the muck and mire of the prima materia, transmute the darkness into light. We fell so that we could experience the nature of duality and material existence. We fell so that when the time came, when the final moment of reintegration and enlightenment was upon us, the reunion with the supreme source would be so sweet.

On a symbolic level, the galactic alignment represents an opportunity for us to reverse this process. By quieting the mind and listening to the voice of the heart, by discovering the place of silence within and entraining one's heartbeat with the galactic pulse of the great mother, the journey begins. In this space of quietude, we can begin to seek out, perceive, and burn away the layers of darkness that surround us physically, mentally, and emotionally. Through the clear light of right action, truth, and integrity, through the power of love and compassion, the essential spark or seed of the Shakti/Shekhina is nourished and begins to germinate and grow within us. The hardened husks of the kelipot and kleshas, the negative thoughts and emotions, the habitual patterns and karmic propensities are melted away.

From a kabbalistic perspective, as this happens, the holy waters begin to flow downward from Keter, the crown chakra, through the central channel of the body into Malkhut, the base chakra and earthly realm of the Shekhina, to make contact with and ignite the Kundalini Shakti, the goddess energy that lies within us. Through this contact with the divine radiance, the Kundalini Shakti is awakened. In a wavelike serpentine rhythmic flow, she rises upward through the swirling vortexes of the chakras and sefirot.[13]

Upward through Yesod she streams, clearing away the last vestiges of our mental and emotional turmoil, the clouds of confusion that envelope us. Into the heart center, Tiferet, she surges, the place of grace, beauty, and compassion. It is here that the Ka, which is on one level the animating force and astral matrix of the physical body and on another level the self or ego with its various personalities, is reunited with the *Ba*, or cosmic soul.[14]

The heart takes wing as the now fully integrated being encounters Da'at, the pineal center, center of the galaxy, and seat of the great mother goddess.[15] Like the son returning to his mother's arms, like the bride coming to the marriage bed, the long years of exile have ended, and the time of union is at hand. Shakti is reunited with Shiva, the Shekhina is reunited with the Lord God, as Isis takes you in her loving arms and brings you home. As your eternal spirit soars from the Earth, to the Sun, to the center of the galaxy, to the divine source, your luminous energy body begins to glow with divine radiance.[16] At the culmination, the omega point of this experience, you are transmuted into a shining being of light, filled with the shimmering resplendent power of divinity. You have now become a fully realized being able to assist all sentient beings on the path of enlightenment. [17]

Traditional teachings throughout the world maintain that in every moment, one has the opportunity to encounter and walk this eternal path of return. Yet, as the alchemists stated, there are certain times that are the most conducive to performing the great alchemical experiment. The galactic alignment in 2012 and the years leading up to it hold the promise and potential of the dawn of a new galactic/lunar age that supplants the toxic, dying throes of the solar age. This collective moment of opportunity is upon us. It is now up to us to perceive, understand, and fully experience its extraordinary transformative power.

The Advent of the Post-Human Geo-Neuron

GEOFF STRAY

*What do the I Ching, the Mayan calendar, the Hopi prophecies,
near-death experiences, and the consciousness of Earth have in
common? According to 2012 expert Geoff Stray, they all concern
the evolution of humankind. One of the foremost collectors
of all things 2012, Stray explores in the following essay a wide
variety of sources and what they have to say about 2012
to help us draw our own conclusions about what is to come.*

In 1971, brothers Terence and Dennis McKenna traveled to the remote Ecuadoran rain forest to sample a shamanic jungle brew that had been heard of by only a few anthropologists. It has now become renowned as the *ayahuasca* brew, named after the vine that is its main ingredient. After a prolonged "contact" experience, the brothers returned to the United States, and Terence set about investigating the clues he had been given in Ecuador. In an ancient Chinese oracle, the I Ching, he would find an encoded "novelty wave" that demonstrates the acceleration of change in the universe toward a point of maximum novelty, when time will end.

The I Ching is a collection of 64 hexagrams, each made up of six-line combinations of yin and yang—broken and solid lines—like positive and negative, or male and female. There are thus 384 lines in the sequence. A complex mathematical procedure of turning all the information in the original King Wen sequence into a graph allowed a study of the times of increased novelty or trends toward habit (events that had already happened before).

To determine where we are on the time wave, the explosion of the bomb at Hiroshima in 1945 was taken as a huge "ingression of novelty" and was lined up with a large novelty peak. Just sixty-four 384-day lunar years later, each consisting of 13 lunar months, the graph hits the baseline—the maximum novelty point that Terence McKenna called the "hyperdimensional object at the end of time." In 1975, the McKenna brothers published their theories in a book called *The Invisible Landscape*, in which they revealed that the termination point of the time wave would occur on November 17, 2012.

NEAR DEATH

On September 17, 1975, Dannion Brinkley was speaking on the telephone during a thunderstorm. A lightning bolt hit the phone line, sending thousands of volts of electricity through his head and body and throwing him near the ceiling of the room. His heart stopped, and at the hospital he was pronounced dead, and a death certificate was issued. He was then covered with a sheet and wheeled toward the morgue. . . . Then, he woke up. He had been clinically dead for 28 minutes. At that time, this was the longest clinically documented near-death experience ever recorded. Brinkley later told how he had gone down a tunnel into a bright light, where he was confronted by a bright silver being and went through a life-recall self-judgment process. He was then taken to meet 13 beings who presented him with 300 scenes from the future, including the nuclear accident at Chernobyl, the breakup of the Soviet Union, the Gulf War, and changes to the Earth's climate. Of 117 major world events shown, 100 of them had become reality by 2002, leaving only 17 still to occur before the last prophecy. In the series, which was as follows, Brinkley said that "between 2004 and 2014, more precisely between 2011 and 2012," there would be "the return of an energy system that existed here a long time ago," and it would culminate in an "electromagnetic polar Earth shift" between 2012 and 2014. The whole process would present humankind with a spiritual opportunity to raise its consciousness.

Also in 1975, José Argüelles published a book called *The Transformative Vision*. A very brief mention on page 305 revealed that 2012 is the year in which the era-length (5,125-year) cycle of the ancient Mayan civilization comes to its conclusion. Argüelles expanded on this in 1987 in *The Mayan Factor*, bringing it to the attention of the world at large. Terence McKenna knew nothing of the Mayan end-point (he updated the 1993 edition of *The Invisible Landscape* to include this)—nor did Brinkley. Since 1987, various people have started announcing 2012 as the end of the world. Christianity has been predicting an imminent Day of Judgment for the last 2,000 years, culminating in the year 2000, when, thankfully, the world yet again failed to end—and the Y2K prophecy of mass computer meltdown didn't happen either. It is easy to get cynical about what seems to be another pseudo-apocalypse. However, the familiar prophecies of doom were all Bible-based, while the origins of the Long Count calendar are 1,500 years before the birth of Christ. Brinkley's prediction suggests that there could be changes for the better, as well as possible extreme climate changes.

OLMECS AND MAYA

The Olmecs, the first civilization known to have inhabited Mesoamerica, are said by archaeologists to have arrived there between 1800 and 1500 B.C. Several Maya scholars now agree that the calendars used by the Maya were developed by the Olmecs in a town called Izapa, which became populated between 1500 and 800 B.C. and which the Maya inhabited from about 250 B.C. According to Mayanist Munro Edmonson, the tzolkin (literally, "count of days") calendar was developed by the seventh century B.C. and consists of 260 days that are a combination of 20 day-signs and 13 numbers. The calendar was used as an almanac, giving each of the 260 days a unique quality that not only influenced the character of those born on that day, but also made performing certain tasks favorable or unfavorable. Today, the descendants of the ancient Maya number about six million, and some, who live in the highlands of Guatemala, have been using this 260-day calendar in an unbroken sequence for thousands of years. When asked why the sacred almanac has 260 days, they reply that it is based on the period of human gestation.

The Long Count calendar is the one that pertains to the 2012 question, and it has a direct relationship to the tzolkin, since it consists of 260 katuns rather than the 260 days of the tzolkin. The Long Count was fully developed by the middle of the fourth century B.C., according to Edmonson, although the oldest stela dating a contemporary historical event shows a Long Count date equivalent to 36 B.C. The Long Count consists of a hierarchy of cycles: 20 kin, or days, make 1 uinal, or 20-day period; 18 uinals make a tun (a 360-day year); 20 tuns make a katun; and 20 katuns make a baktun. After 13 baktuns have passed, a creation date is reached, which is written 13.0.0.0.0. The last creation date occurred on the Gregorian calendar equivalent of August 11, 3114 B.C., and this is the "base-date" of the 13-baktun cycle, from which all dates inscribed on ste-lae were counted. The next time the Long Count date reaches 13.0.0.0.0 will be on December 21, 2012.

PRECESSION

Mayan scholar John Major Jenkins has made a strong case that the Long Count calendar was actually a device for tracking the cycle of precession. This is the 26,000-year cycle in which Earth's axis, which is 23.5 degrees off "vertical," rotates; it is measured by the movement of constellations against the fixed positions of the rising equinox or solstice Sun. The winter-solstice 2012 termination point of the 13-baktun cycle indicates a 36-year window (1980–2016) in which the winter-solstice Sun rises on the galactic equator. Astronomers variously put the exact halfway point of this process at 1998 or 1999. The Olmecs and Maya targeted it at 2012. Some commentators have said that the Maya were thus 13–14 years out with their end point, which is still quite impressive, considering it was calculated more than 2,000 years ago by a nontechnological society. However, as we shall see, there are reasons for considering that the year 2012 is a significant year in the galactic alignment process and that it was deliberately targeted.

The *Popol Vuh* myth of the Maya tells of three previous ages, or Suns, that each ended in destruction. This is reflected in the four worlds of Hopi mythology (see page 336). However, there is also a Tzutujil Maya myth, recounted by Martin Prechtel, that indicates we are now in the Fifth Sun. Various Aztec myths also describe four previous ages or Suns and a current Fifth Sun, but they differ as to what kind of destruction ended which Sun (wind, a rain of lava, an eclipse with monsters and flooding—the current Sun is expected to end in an earthquake). These five Suns are depicted at the center of the Aztec Sun stone, which is a 12-foot diameter engraved piece of basalt that also shows the Aztec versions of the 20 day-signs of the 260-day sacred almanac. There are two Mayan murals called Murals of the Four Suns—one at Palenque and the other at Tonina, both in the Chiapas state of Mexico. They show a central skull-like deity, with four "tassles" or decapitated heads below, each representing a new rebirth of the Sun—the new Sun, born on the creation date 13.0.0.0.0. If the central controlling skull is compared with the same one on the Aztec Sun stone, it suggests that the current Sun is actually the Fifth Sun, or World, rather than the Fourth. In fact, Jenkins makes a case that the current 13-baktun cycle is the fifth in a series. Each Sun, or 13-baktun cycle, thus consists of 5,200 tuns, and the series of five Suns would total to 26,000 tuns, or one cycle of precession.

THE PETRIFIED CHRONOMETER

The Toltec people of northern Mexico, like the Maya, used the 260-day almanac and also a 365-day calendar known to the Maya as the *haab*. Each day had a name in the 260-day calendar and a name in the 365-day calendar, but it took exactly 73 tzolkins and 52 haabs for the combination to repeat itself. This period is known as a calendar round, and for every calendar round, the Toltecs held a ceremony called a New Fire Ceremony six months before the day of the zenith passage of the Sun (when it is directly overhead). John Major Jenkins has shown that this New Fire Ceremony was originally a method of tracking the cycle of precession. The Toltec knowledge was bequeathed to the Aztecs, and it is recorded that exactly six months before the New Fire Ceremony, the priests would go to the top of a hill to see if the Pleiades reached the zenith at midnight. If the constellation reached the zenith, it meant that the world would not end. They would then know exactly how close the Pleiades would be to the zenith Sun six months later, when the stars were not visible due to the daylight.

When the Toltecs moved to the Yucatán Peninsula in southern Mexico, it seems that they merged their zenith cosmology with the Mayan system and encoded the result into the Pyramid of Kukulcan at Chichen Itza. The pyramid has 91 steps on each of its four sides, totaling 364, plus the top step: 365 altogether. This is a clear sign of a calendrical meaning. Every year, on the spring equinox, the afternoon Sun causes a shadow play on the pyramid, so it appears that a huge snake is descending the pyramid from the sky. The effect is enhanced by stone snake heads at the bottom of one staircase. The *Crotalus durissus* rattlesnake, whose zigzag pyramid pattern has heavily influenced Mayan art and architecture, also has a marking close to its rattle that is identical to the Mayan Ahaw glyph, which is associated with the Sun. The rattle itself is called *tzab* in the Yucatec Mayan language, and this is also the word for the Pleiades constellation. Jenkins argues that Kulkulcan, the Mayan name for Quetzalcoatl, the plumed serpent of the Toltecs and Aztecs, is thus a symbol for the Sun-Pleiades-zenith conjunction, and sure enough, we are now at the beginning of a 360-year window when the Pleiades conjuncts the zenith Sun directly over the Pyramid of Kukulcan. At the start of the 360-year window is 2012, and on May 20, 2012, 60 days after the equinox snake shadow play, the Pleiades-zenith-Sun conjunction will occur on the same day as a solar eclipse, on the day 10 Chicchan in the unbroken tzolkin count of the highland Maya. *Chicchan* means snake, and the snake rattle is sounding our wake-up call.

THE HOPI EMERGENCE

According to Frank Waters's *Book of the Hopi*, the Native American Hopi say that we are in the fourth of seven eras—the Fourth World—and that we are approaching the transition to the Fifth World. The First World ended by fire; the Second World ended when the Earth "teetered off balance," causing floods and an ice age; and the Third World ended in a flood. There will be a "great purification" just before the start of the Fifth World. The transition is called an emergence, and every year, in November, the Hopi have a ceremony called *Wuwuchim*, held in an underground chamber called a *kiva*, in which they reenact the emergence, which is seen as a birth process, and "initiates undergo a spiritual rebirth." There is a hole in the floor of the kiva, symbolizing the previous ascent, or emergence, from the Third World to the Fourth World. A ladder leading up and out of the kiva represents the forthcoming emergence from the Fourth World into the Fifth World. The ceremony includes a New Fire Ceremony and culminates at midnight, when the Pleiades constellation is overhead. This ceremony shows a direct connection with the Toltec (and later Aztec) New Fire Ceremony, which Jenkins has decoded as a precession-tracking method, as encoded in the Pyramid of Kukulcan "stone alarm clock," with its 2012 rattlesnake alarm.

Moses Shongo, the last of the traditional Seneca medicine men, who died in 1925, prepared his granddaughter, Twylah Nitsch, to carry on his teachings. They included a prophecy that there would be "a 25-year period of purification lasting until the year 2012 or so, during which the Earth will purge itself." Another Native American, a Pueblo Indian from northern New Mexico, called Speaking Wind, has recently confirmed that the Fifth World will start in December 2012. In the 1980s, the Paqos, priests of the Q'ero tribe living high in the Peruvian Andes, announced that their Pachakuti formula—the turning over of the world—would start in 1990 and last 22 years. Thus, 2012 will bring the start of a golden age called *Taripay Pacha*, when the upper world, lower world, and everyday world will unite. *Pachamama* means mother earth, and we see here another term that means world, as in Earth and World era. The Q'ero say that the Taripay Pacha will be "the age of meeting ourselves again." However, Willaru Huayta, a Peruvian spiritual messenger from the Quechua nation, says that in 2013, a huge magnetic asteroid will pass close to Earth, and only those who have conquered their egos will survive. This more catastrophic version is echoed by the Zulu shaman Credo Muttwa, who says that in the year of the red bull (2012), a star with a very long tail called Mu-sho-sho-no-no will return. The last time it came, it brought a deluge and turned the world upside down, so that "the Sun rose in the south and set in the north."

GEOMAGNETIC REVERSAL

We can see hints here that the "turning upside down of the Earth" may actually refer to a shift in the magnetic poles, accompanied by a comet or asteroid sighting and events on Earth such as flooding (or perhaps seismic and volcanic activity, depending on regional variations). The 5,125-year 13-baktun cycle is exactly one-eighth of the 41,000-year period of the "variation of obliquity" cycle (change in the angle of axis tilt) that is used to calculate ice-melt cycles. The baktun itself is 400 years long, which is the period of rotation of the Earth's core. This invites the speculation that a geomagnetic reversal is in the cards for 2012. By many accounts, the Earth is overdue for a magnetic reversal, and although most geologists would say a reversal takes thousands of years, the Steens Mountain volcanic record indicates that the field can move at up to 6 degrees per day, meaning that a reversal could happen in only 30 days.

However, from New Zealand comes a creation myth that suggests a reinterpretation of the catastrophic angle may be in order. The Maori creation myth, which describes the original separation of Earth and sky, says that one day, when humankind is distracted, the Earth and sky will crash back together, destroying everything. The term for the event translates as "the curtain will fall," which has been interpreted as the end of the world. However, when a young Maori recently interviewed the elders, it transpired that not only do the elders predict this event for the year 2012, but also that the Maori language has evolved since the myth originated, and the word for "curtain" originally meant "veil"—the same word used when people have died and gone "behind the veil"—and the word for "fall" originally meant "dissolve." In fact, the prophecy should say "the veil will dissolve."

THE SPIRITUAL WEDDING

It seems that "the end of the world" actually means the end of the current era, and there are at least three more to go, according to Hopi mythology. Archaeologist Laurette Sejourné says that the Aztecs misunderstood the Toltec religion, which was based on the concept of rebirth. The symbol of rebirth is the snake, because it sheds its skin, and the Aztec practice of skinning captives and wearing their skins was a result of this misunderstanding. Quetzalcoatl, the plumed serpent, who is represented on the "stone alarm clock" at Chichen Itza, represents a union of Earth (snake) and sky (feathers). The explanation comes from the work of Mary Scott, who is versed in the Hindu lore of Kundalini yoga. The Kundalini is said to be a fire snake, coiled in three and a half coils in the base chakra (one of seven power zones along the human spine). It is coiled around the Shiva lingham—the male line of force that comes down from the sky. Kundalini, says Scott, is a force that comes up from the Earth. She also notes that there are several findings by dowser Guy Underwood, who wrote *Pattern of the Past*, that confirm this connection. For example, the geospirals and blind springs found by dowsers are always in coils that are multiples of three and a half. Kundalini yoga is the practice of awakening the snake and allowing it to rise up the spine until it reaches the highest chakra, the crown chakra, where the meeting of Shiva and Shakti leads to enlightenment and the trance of Samadhi—ecstasy—the annihilation of the ego and the end of duality. Could this be the meeting of the Maori sky and earth gods? Could it be the merging of the Incan overworld and underworld into the everyday world? Could it even be the descent of Christ and the ascent of the bride of Christ?

The original Toltec religion of rebirth associated with Quetzalcoatl has now been revealed in a long-lost Aztec codex called *Pyramid of Fire* (rediscovered and published by Jenkins). The codex actually explains that the Aztecs did misunderstand the metaphors, thinking that the sacrifice of the ego and the release of energy from the heart were about human sacrifice. It says that Quetzalcoatl is "the man that achieves God" and that everyone has his or her own serpent, which is the energy of Tonatiuh, the Sun. "And in this serpent sleeps consciousness; in this serpent is hidden divinity. From this serpent his wings will grow." In Aztec myth, Quetzalcoatl is associated with Venus; Quetzalcoatl sacrificed himself and became Venus. The eight-day disappearance of Venus at inferior conjunction was associated with the eight days that Quetzalcoatl spent in the underworld before reappearing as Venus. The most brilliant apparition of Venus is when it appears close to the Sun, rising just before the Sun as the morning star, just after this eight-day disappearance. So the movement of Venus encodes the rebirth concept. As we have seen via the Pyramid of Kukulcan, there is a connection between Quetzalcoatl/Kukulcan and 2012.

It seems that the Aztecs may have been expecting the return of Quetzalcoatl between two Venus transits (which occur during the eight-day disappearance), as there were a pair of transits (which are always exactly eight haabs of 365 days apart) in 1518 and 1526. Cortez arrived in 1519 and was mistaken for Quetzalcoatl (the Aztecs, again, taking the myths too literally). We are in between another pair of Venus transits: one occurred on June 8, 2004, and the next is due on June 6, 2012. Both these transits occur on an Ik day in the unbroken tzolkin count of the highland Maya. Ik, in the Aztec version of the 260-day almanac, was Ehecatl, a form of Quetzalcoatl. So, the Venus transits actually occur on the day of Venus. Just before the next transit, the Pleiades-zenith Sun conjunction will occur over the Pyramid of Kukulcan, on snake day, confirming that 2012 is the correct date for the return of Quetzalcoatl.

PLASMATIC IMMERSION

The "spiritual opportunity for humankind," spoken of by Dannion Brinkley, can be confirmed as a time in or very close to 2012, when something will trigger a mass Kundalini experience in humankind. What could the trigger be? Some energetic effect of the galactic alignment process itself? The Russian biochemist Simon Shnoll has established that human biochemistry (and neurochemistry) is affected by the Earth's orientation to the stars and by the sunspot cycle—due to reach solar maximum in 2012. Another scientific paper, published by James Spotiswoode, indicates that psychic ability in humans increases exponentially when the galactic center is rising on the horizon.

Of all the theories various researchers have put forward to explain 2012, including comets, rogue planets, asteroids, galactic core explosions, mega-sunspot cycles, obliquity harmonics, unusual conjunctions, and Milankovitch cycles, there is one theory that is scientifically valid and that fits Dannion Brinkley's scenario very well. Alexey Dmitriev, a Russian geologist, has analyzed the increasing changes in Earth's weather and seismic phenomena, plus recent changes to the geomagnetic field. He has also taken into account solar changes and changes in the weather and magnetic fields of all the other planets in the solar system. A tenfold increase to the layer of deflected interstellar plasma on the heliosphere boundary indicates that the solar system is headed into an area of magnetized plasma that, Dmitriev says, is causing these changes in weather and other patterns. An increasing incidence of plasma balls in the atmosphere is a sign of the imminent "transformation of Earth" involving an interaction with processes beyond the three-dimensional world, says Dmitriev.

Incredibly, Brinkley is not the only person who has returned from a near-death experience (NDE) with predictions of Earth changes and consciousness changes associated with 2012. In 1969, Ken Kalb had an NDE with visions that resulted in his writing a book called *The Grand Catharis*, about consciousness changes leading to 2012—a period he saw as Earth's own NDE—and a global rebirth. In 1992, Cassandra Musgrave had an NDE after drowning during a water-skiing accident and foresaw an acceleration of Earth changes leading up to 2012. Pam Reynolds had an NDE while dying from a brain tumor and returned to tell of Earth changes, a consciousness change, plus a physics breakthrough in 2012. Phylis Atwater had three NDEs plus Kundalini experiences. She then became an NDE researcher and studied child NDEs, and the children's predictions again culminate around 2012. She went on to study the so-called Indigo Children—the gifted and psychic children who today are being born in greater numbers than ever before.

The psychic children, she says, show the same qualities as those who have returned from NDEs, and they are the forerunners of a new evolutionary stage.

Dr. Kenneth Ring, who completed a huge study of NDEs called The Omega Project, says these permanent changes consist of improved self-esteem, more altruism, a less materialistic approach to life, a more broad-minded and tolerant approach to spiritual and religious beliefs, and many instances of telepathy, clairvoyance, precognition, and other psychic abilities. He has observed that due to advances in revival technology, more and more people are experiencing NDEs, and he proposes, citing Rupert Sheldrake's formative causation theory, that they will reach a critical mass. This is the moment of morphic resonance: when the critical mass is reached, consciousness changes become species-wide.

THE TORTUGUERO PROPHECY

In April 2006, awareness of an important inscription at the classic Mayan site of Tortuguero entered the public domain. Mayanists have been aware of Monument 6 for about three decades, but it was a direct question from anthropologist Robert Sitler on an online specialist discussion group that produced a new translation from epigrapher Dave Stuart. The monument is damaged and incomplete, but this much has been translated: "The Thirteenth Bak'tun will be finished [on] Four Ahaw, the Third of K'ank'in. [?] will occur. [It will be] the descent [?] of the Nine Support [?] God[s] to the [?]."

The cross-referencing of the Long Count, tzolkin, and haab calendars unequivocally identifies this date as December 21, 2012. This late Classic era inscription echoes the prophecies of the post-conquest books of Chilam Balam (the jaguar priest). The Long Count had fallen out of use by the time the Spanish invaded, and instead of the 13-baktun cycle (that consisted of 260 katuns), the Maya were using a 13-katun cycle (that consisted of 260 tuns). Mayan scholars call this the Short Count. In her translation of the Chilam Balam of Tizimin, Maud Makemson makes a case that some of the prophecies that seem to relate to the end of the Short Count must originally have applied to the end of the 13-baktun cycle of the Long Count. Like the Tortuguero prophecy, the Chilam Balam prophecies foresee a return of the gods at the end of the 13-baktun cycle. Kukulcan is predicted to return in katun 4 Ahaw. This is the 20-year period we are now in, up to 2012—the same 20-year period that emerges in the NDE prophecies. Makemson's translation says,

> Presently Baktun 13 shall come sailing . . . then the god will come to visit his little ones. Perhaps "After Death" will be the subject of his discourse . . . in the final days of tying up the bundle of the 13 [baktuns] on 4 Ahaw. . . . I recount to you the words of the true gods, when they shall come. . . . Day shall be turned upside down . . . stones shall descend and heaven and earth shall be universally consumed by fire . . . the Dead shall live! Dying from old age they shall immediately ascend into heaven.[1]

The return of Kukulcan suggests a mass Kundalini experience, while the other prophecies speak of Earth changes and an apparent mass NDE for humanity, perhaps propelling us into a post-human state.

THIRD EYE OPENING

The prophecies of 2012 are not just from those who have experienced NDEs. There are similar reports from people who have undergone out-of-body experiences (OOBEs), lucid dreams, remote viewing, meditative trance states, hypnotic progression, and even alien abduction ("encounter") experiences. It seems that a clue may lie in human neurochemistry. Terence McKenna accessed the information about 2012 when using an entheogenic mushroom, and the Maya used the same mushroom, which contains psilocybin—a compound closely related to dimethyltryptamine (DMT). Dr. Rick Strassman has recently confirmed that at birth, death, and during mystical experiences, including encounter experiences, the pineal gland (which is very sensitive to changes in the geomagnetic field), at the center of the human brain, secretes DMT into the blood. Dr. Callaway of the University of Kuopio, Finland, has supported Strassman's findings and "conclusively demonstrated that NDEs, OOBEs, and death itself are based on pinoline, DMT, and 5-meO-DMT release," and also that lucid dreams are induced by DMT. Pinoline and 5-meO-DMT are two more hallucinogens internally produced by the pineal gland. Pinoline is also in Virola—one of the variable ingredients of the ayahuasca brew. The ayahuasca vine itself contains a very similar molecule to pinoline, called harmine. In fact, harmine was actually called telepathine when it was first extracted from the vine, due to its effects; then it was found that the same alkaloid had already been extracted from Syrian rue (*Peganum harmala*) and named harmine.

Could it be that our entry into the magnetized plasma band will cause Earth weather changes and a geomagnetic reversal, which will trigger the pineal gland into DMT and pinoline production, leading to Kundalini and out-of-body experiences and increasing incidents of telepathy, while we develop our ability to see beyond the veil into the invisible landscape?

Perhaps the Mayan shamans and the medicine men and priests of North and South America were able to activate the secretion of DMT and pinoline (or achieve the same result using psilocybin or ayahuasca—the jungle brew containing DMT and pinoline/telepathine) to access some sort of global megamind—the developing consciousness of the Earth mother. After all, the 13-baktun cycle is a 260-unit cycle and thus a macroscopic version of the 260-day sacred tzolkin calendar, which is based on the period of human gestation. Hence, the 13-baktun cycle may measure a planetary gestation period of 5,125 years, which is the length of human civilization's history. According to Peter Russell, there are 10^{10} molecules in a human neuron, and 10^{10} neurons in the human neocortex of the brain. The population of the Earth is currently over 10^9 and heading for 10^{10}. When the critical number is approached, the parts lose their individual identities ("egos") and become components of a larger whole. Thus the Earth is about to give birth to its own human neocortex, as we become human neurons in a global brain, with telepathic synapses.

PART FOUR

A New Humanity: Evolution
Toward 2012 and Beyond

How the Snake Sheds Its Skin

A TANTRIC PATH TO GLOBAL TRANSFORMATION

DANIEL PINCHBECK

Daniel Pinchbeck explains, "Shamanism is a technology for exploring non-ordinary states of consciousness in order to accomplish specific purposes: healing, divination, and communication with the spirit realm." In his book 2012: The Return of Quetzalcoatl, *Pinchbeck takes the role of a modern shaman and embarks on a journey of exploration into what he calls the "extravagant thought experiment" that revolves around 2012, engaging in exhaustive research, including the use of psychedelics. In his book, he investigates Burning Man, Stonehenge, crop circles, aliens, and many other phenomena, ultimately becoming a journalistic expert on 2012 and what the next decade has in store for us. In the following essay, he gives readers an overview of his most current realizations related to 2012 and describes how we might use a tantric path to escape destruction and pursue a new age of consciousness.*

That we could interrupt our current suicide march and institute a harmonic and peaceful planetary civilization within the next few years, embrace a non-dual realization of the universe, and institute practices of deep ecology and collective psychic ritual may seem far-fetched to some. The history of human thought, however, reveals an extraordinary tendency for ideas that at first seem radical, absurd, or beyond comprehension to quickly become commonplace, even truisms, leading to new conditions and transformed social structures. Before the eighteenth century, for example, people knew about lightning, but nobody had any idea that electricity could be shaped to human purposes and made into a transformative energy for the world. Once someone learned this trick, we engineered industrial civilization and changed the entire surface of the Earth in less than two centuries—not even a blink of evolutionary time. If we did it then, we could do it again, under different conditions, with different goals, on a far more concentrated timescale.

This is an unprecedented moment for humanity. The material progress of the last centuries has slammed against the limits of the biosphere, putting our immediate future in jeopardy. Equally surprising to the modern mind, many indigenous cultures around the world recognize this time through their prophecies and myths and may possess insight into its deeper meaning. For these cultures, we are transitioning between World Ages; this shift not only involves changes in the physical body of the Earth, but also a transformation of human consciousness—a regenerative cycle of initiatory death and rebirth, leading to a higher level of manifestation.

Knowledge of the importance of this time appears embedded in the harmonic gyres and fractal intricacies of the Mayan calendar, which completes its Great Cycle of 13 baktuns on December 21, 2012. The Classic-era Maya created a sophisticated culture on the Yucatán Peninsula in Mexico and in Guatemala, before mysteriously vanishing in the ninth century A.D. Despite the increasingly glaring deficiencies of our civilization, most modern people find it far-fetched that a nontechnological and myth-based society, such as the Classic-era

Maya, might have developed a different form of knowledge that, in crucial respects, was more advanced than our own. Contemporary scholars outside academia have advanced the hypothesis that the Maya attained a precise realization of cyclical patterns in human development, linked to astronomical cycles. In the works of John Major Jenkins, Carl Johan Calleman, and José Argüelles, this argument has been developed with compelling sophistication.

In my two books, *Breaking Open the Head* and *2012: The Return of Quetzalcoatl*, I first stumbled upon and then rigorously pursued the puzzle pieces that indicated our modern world was constructed on an "irrational rationality," a reductive empiricism requiring a rigorous suppression of psychic, intuitive, and shamanic dimensions of being. While earlier civilizations and shamanic cultures may have been incapable of constructing a cell phone or an iPod, they did possess a participatory and holistic relationship to the universe, encapsulated in the alchemical catchphrase, "as above, so below." They reified their wisdom traditions in monumental structures, such as the stone circles of England or the Mayan pyramids, which were carefully attuned astronomical instruments and ritual centers. They utilized non-ordinary states of consciousness, accessed through psychedelic plants and other means, in an applied shamanic science.

As huge populations across India and China grasp Western-style material comforts for the first time, an elite subset in Europe and the United States has shifted its focus to ecological sustainability, new and ancient alternatives to modern medicine, and spiritual disciplines. Today, 40-plus years after the first irruption of shamanic consciousness into the modern psyche during the curtailed psychedelic era of the 1960s, we find increasing numbers of people recovering elements of these lost traditions, exploring shamanic practices, tantra, yoga, and other ancient and indigenous techniques for attaining intensified states of awareness. Those of us at the vanguard of this transformational process are learning that the participatory rituals and spiritual wisdom of indigenous and myth-based cultures hold tremendous significance for our lives. We have embarked upon a new phase of the initiatory journey begun a generation ago—with the opportunity to avoid the tactical mistakes, strident statements, and polarizations of the past.

Whatever the Maya may or may not have known about this time, we are confronting the most critical juncture for our species since our origins. During the winter of 2006–2007 alone, climate change appears to have passed a tipping point—in New York City, in early December, temperatures were 15 degrees higher, on average, than they were the previous year. Flowers that usually do not appear until March bloomed in Central Park. These are only the most tangible indications of the traumatically inconvenient truth of our current predicament—a truth that is still avoided and strenuously suppressed by government functionaries and corporate CEOs.

The 400-year paradigm of modern civilization, as currently conceived, has reached the end of its line. Without a radical reorganization and rapid transformation of global practices, there is a high probability that the human species will soon crash and burn, as so many species have before us. While climate change accelerates, polar ice caps melt, and rain forests are cut down, more than 25 percent of all species will be extinct within the next 30 years—the global disappearance of frogs and amphibians indicates a future that may be inhospitable to many forms of life, including our own. From a purely materialistic and rational analysis, the next few years represent a decision point for humanity, and most of the U.S. population, as well as millions upon millions outside of this continent, are sleepwalking through the supermarket aisles, hypnotized by programmed infotainment, assuming that the present situation can continue without a radical rupture from what has gone before.

While paralysis and avoidance are common reactions to the crisis looming over us, an alternative vision is possible. What appears as unavoidable chaos and collapse is actually a natural process accompanying an evolutionary advance in human consciousness. What we are destined to experience within our lifetimes is not the end of the world, but the birth of humanity's higher mind. We are not helpless spectators in this drama, but powerful actors, with an opportunity to directly influence the results.

In what follows, I will sketch out the bare rudiments of this alternative paradigm, based upon psychic realities as well as material ones. Following this, I will briefly describe some developments and techniques that, if implemented in time, could have a positive influence on the course of events. By this hypothesis, a desirable outcome can be realized if an elite vanguard overcomes all obstacles and prior conditioning to attain an intensified awareness of the situation, and then works efficiently and collaboratively to propagate this new paradigm across the Earth.

In *2012: The Return of Quetzalcoatl,* I theorized that we are currently in the process of shifting from one form of consciousness, one way of being, to another. As a by-product of developing empirical and scientific thought, the modern mind trapped itself in a mechanized conception of the universe, a limited realization of time, and a denial of spirit and intuition. Quetzalcoatl—the feathered serpent of Mesoamerican myth—symbolizes the meeting of bird and snake, or Heaven and Earth. I propose that this archetype also represents the integration of rational and empirical thought with shamanic, intuitive, and esoteric knowledge. I compiled evidence for this thesis from many fields and diverse philosophical traditions, isolating three trends that seemed the most crucial indicators. The first, as described above, is the ecological crisis, including accelerated climate change, species extinction, and depletion of resources. The second is the rapid development of new technology, especially communication and media technologies, ranging from social networks to cell phones, as well as the potentially disruptive capabilities of nanotechnology and other future, increasingly plausible advances.

The appearance of new media and communications tools accompany profound changes in the human psyche, along with new forms of social organization. This is obvious when we look at past examples, such as spoken language with tribal culture, written language with hierarchic civilization, and the invention of the printing press with the emergence of mass democracy. Currently, the use of cell phones, the Internet, and social networks is reformatting human consciousness in ways both trivial and profound. Because we find ourselves enmeshed in these developments, adapting new habitual and subliminal patterns, it is difficult to analyze how they affect us.

My hypothesis is that these new tools impact our inner experience of personal identity and reformat our sense of self. Through the last centuries, the modern Western person developed a strong sense of individual autonomy, alienation from the natural world, and existential isolation. The new communication tools soften and diffuse the boundaries of the ego. Increasingly, with our networked lifestyles, we experience our identity as contextual, fluid, and relational, rather than a separate entity that is fixed and permanent. Academics have long argued that identity is a social construct, but this phenomenological shift makes it explicit. Our individual sense of self depends upon greater networks of connectivity, while the pervasive media, surrounding us like a second skin, continually reshapes collective viewpoints, personal boundaries, and moral structures. Through numbing repetition, the virtual sphere of electronic media can make actions that were once reprehensible, such as government-sanctioned torture, socially acceptable. As media and communication tools unify the collective mind, mass delusions become a greater danger. Yet, a new social movement or a more authentic awareness could also spread quickly, just as a flock of birds responds to danger with efficient coordination.

The third trend is harder for many to grasp and impossible to quantify, as it can only be realized subjectively. I propose that an evolution is also taking place in the nature of the psyche, and that one expression of this is an accelerating surge in synchronicities, telepathic hints, and psychic as well as psychophysical episodes of all types. For those who are paying attention to this subtler level of phenomena, material reality seems less dense and, in quickening degrees, more psychically responsive.

The veil between matter and consciousness is becoming thinner and more permeable. This is evident when we consider the rapid evolution of technology. Technology allows us to turn our "thought forms" into fully realized manifestations, at an ever-increasing rate of speed. This is literally happening, and it is undeniable. Fifty years ago, you might have imagined a complex sculptural form or musical composition and been able to execute it after years of effort. Today, assuming you have the resources, digital tools allow you to produce any form or soundscape in a matter of hours. The extensions of technology allow the imagination to realize its projections with greater accuracy and speed, drawing the psychic and the physical into an ever-more-intimate relationship.

Beyond these overtly material expressions, the shift in the nature of the psyche reveals itself in a transformed relationship to reality. As we integrate an awareness of the archetypal domain, or the higher-dimensional, space-time continuum defined by quantum physics, the reality surrounding us takes on a different valence. We begin to experience the world as a mandalic gyre, a Gnostic parable, in which our being is enmeshed, rather than an arbitrary zone of meaningless accidents. Unlike material phenomena that can be quantified, this intensifying of psychic energy only becomes available to us when we focus our consciousness on it. To borrow a term from scientist Rupert Sheldrake, the "morphogenetic field" for psychic capabilities increases in force as more people interact with it, feeding it energy with their attention and awareness.

In the twentieth century, psychoanalyst Carl Jung realized "the reality of the psyche," the interconnection of mind and world, the psychic and the physical, expressed through synchronicities and other occult correspondences. Jung understood that a deep transformation of the human psyche was underway. According to Jungian theorist Edward Edinger, we are currently experiencing the "archetype of the Apocalypse," a term that has familiar destructive connotations but that is also defined as "revealing" or "uncovering." As a negative archetype, it represents the smashing of previous forms of thought and ways of being; as a positive archetype, it represents a momentous event—"the coming of the Self into conscious realization." In the Jungian model, the "Self" encompasses the complete expression of our being, including both conscious and unconscious elements. Our limited egos dread the coming of the Self, yet paradoxically yearn to incarnate it. As Jung described in *Answer to Job* and other writings, the archetype of Christ and the Second Coming represent the fusion of the ego and the Self, a process that takes place within historical time yet is, paradoxically, atemporal. As Christ expresses this paradox, "The hour is coming, and now is."

If the end of the Mayan calendar signals the completion of this archetypal process in historical time, we can begin to appreciate the momentous psychic distance we will travel in the next few years. Today, most of us identify with our individual path and ego-defined goals. As we complete this apocalyptic passage, we will conceive ourselves, increasingly, as fractal expressions of a unified field of consciousness and sentient aspects of a planetary ecology—the Gaian mind— that is continually changed by our actions, and even our thoughts. When the collective shares this understanding, we will move beyond the threshold of history, with its narrative of wars, conquests, and competing belief systems. Humanity will unite as a global tribe, ensuring all of its members equal rights and equal status in a planetary community. Having resolved internecine discord, we will be ready to make our entry into galactic civilization, where we will discover new myths, new companions, and new dimensions of being.

It seems likely that such a profound shift in the collective psyche can only take place as a response to a series of traumatic events, much like the contractions and convulsions accompanying birth. The most comprehensive description I have found of how this process may unfold over the next few years is presented in Christopher Bache's excellent book *Dark Night, Early Dawn*, an extension of Stanislav Grof's theoretical work on non-ordinary states of consciousness and the perinatal matrices. In a series of visions that Bache received during LSD therapy sessions, he foresaw a drastic meltdown of society, due to environmental cataclysm, and then a recrystallization into a planetary culture.

"The events that had overtaken Earth were of such scope that no one could insulate themselves from them," Bache writes. "The level of alarm grew in the species field until eventually everyone was forced into the melting pot of mere survival." New collaborative social structures and expansive family models developed in response to the peril. "It was as if the eco-crisis had myelinized connections in the species mind, allowing new and deeper levels of self-awareness to spring into being." He recounts an extraordinary compression of events: "The pace of the past was irrelevant to the pace of the future. The new forms that were emerging were not temporary fluctuations but permanent psychological and social structures that marked the next evolutionary step in our long journey toward self-activated awareness."[1] According to the time frame developed by Carl Johan Calleman in his book *The Mayan Calendar and the Transformation of Consciousness,* the years 2008–2010 will be the crucible during which the new form of human consciousness emerges as the old structures collapse.

Of course, this prophetic paradigm is only theoretical until events actually come to pass. Each of us will decide for him- or herself whether there is validity to what is presented here and how such a hypothesis will influence his or her activity. If one takes it seriously, as I do, then one is faced with a number of fascinating paradoxes and subtle nuances as the precipice draws near. One basic reaction is to consider escaping the technocratic nightmare and relocating to a rural community on high ground, learning to grow food, and regaining the basic skills that modern society forfeited. While this may be the best option for some people, for many of us, it is out of the question due to family and financial obligations. Personally, my intuition is that any hasty action, motivated by fear, is likely to backfire. Those of us with foreknowledge have a tremendous opportunity—and, perhaps, a responsibility—to reorient the collective consciousness in this interim period. It may be possible to begin constructing those "permanent psychological and social structures" of the future before the breakdown begins. The dissemination of a deeper awareness, along with tools for addressing our immediate situation, could dramatically reduce the approaching crises' destructive repercussions envisioned by Bache and Calleman, dystopian novelists like J. G. Ballard and Cormac McCarthy, and many others.

Despite its threatening material aspects, the basis of this transition between world ages is our psychic reality. If it is the case that mind and matter are increasingly interpenetrating as we approach the 2012 phase-shift, then our level of consciousness and the power of our intention will determine how events play out for us and upon the greater world stage. If we treat this transition as an initiatory ordeal, we will embrace the paradox of becoming increasingly relaxed, open, cheerful, clear, and compassionate as events take their course.

The inner attitude required for this challenge evokes the philosophy of tantra, as defined by the scholar Georg Feuerstein: "The formula 'samsara equals nirvana' implies a total cognitive shift by which the phenomenal world is rendered transparent through superior wisdom," he writes. "No longer are things seen as being strictly separated from one another, as if they were insular realities in themselves, but everything is seen together, understood together, and lived together. Whatever distinctions there may be, these are variations or manifestations of and within the self-same Being." From the perspective of tantra, which fuses the radical techniques of alchemy and shamanism with Vedantic and Buddhist non-dual philosophy, reality is a matter of what we choose to perceive—despite any appearance to the contrary. If samsara is nirvana, then everything is always happening just as it is meant to unfold.

Through tantric techniques, the practicioner recognizes any "other," whether cremated corpse or gorgeous dakini, as an aspect of the Self. When we hold this view, we find there is no problem at all, no crisis requiring some anxious and urgent response. In fact, fear-based reactions can only exacerbate the apparent negative aspects of our current situation. We best serve the transformational process—for ourselves and for the world—by attaining a level of being that is without judgment, anxiety, or negative projections.

Our ability to help the world and heal humanity's traumas depend upon the inner work we have done to master ourselves and attain equanimity of mind, or what Buddhism defines as nonattachment. Nonattachment is subtly, yet profoundly, distinct from detachment. While detachment implies glacial remove and lack of emotion, a nonattached person still experiences the full range of their feelings, their pain, and their joy, but they do not identify with these emotional states. The change required of us is not an unfelt, intellectual shift to some "spiritual" or psychic perspective, but a fully embodied and intimately personal process. We are being called upon to open our hearts, as well as our minds, to the radiant flame of transformation. For those who are willing to take up this challenge, the rewards will be far greater than the sacrifices.

We confront paradoxical difficulties in envisioning and then establishing interim systems to support a large-scale transformation of human society in an accelerated time frame. While "sustainability" has become a trendy buzzword, we lack a strategy for creating a truly sustainable global culture with the current level of population. We find fragments of such a plan in the work of many visionaries, ecologists, radical economists, and design scientists; it should be possible to assemble these scattered projects into a holistic model. However, any solution-based approach will have to remain nonprogrammatic, open to change and indeterminacy. As human consciousness is making a phase-shift to steadier states of mind, our transitional efforts will reflect the awkwardness and uncertainty of this process.

Since we have embarked upon the second stage of the initiatory journey for the modern psyche that was begun in the 1960s, we can expect the typical cycle of shamanic dismemberment and spiritual death-and-rebirth on an individual and collective level—part of the acceptance and integration of deeper levels of the psyche that Jungians call individuation. Over the last decades, those segments of the population with a regressive, authoritarian mind-set—bypassing the individuation process entirely through hypocritical belief systems and an atavistic denial of personal responsibility—organized effectively to exert a baleful influence on public discourse and policy. By contrast, the progressive sector remains uncoordinated and disempowered, hampered by psychological complexes leading to dysfunctional behavior. It would seem that at this point in time, the Left—or at least the majority of its pivotal figures and power brokers—have not resolved their individuation crisis. Rather than overcoming divisiveness, nongovernmental organizations (NGOs), nonprofits, and like-minded companies create allegiances and sentimental attachments to a particular cause or entity. This works against an efficient coordination of forces, on a larger scale, to achieve a collective goal. In many cases, progressive campaigns address particular symptoms but ignore the root cause of the disease. A more productive attitude might echo Nietzsche's definition of the Superman, who combines "the mind of Caesar" with "the soul of Christ."

While U.S. prestige has waned, our popular culture remains a global force. A revival of the transcendentalist impulse in American life could incite a spiritual revolution within our society. If that were to occur, the extraordinary power of our electronic media—which currently functions, for the most part, as a control mechanism, holding the mass consciousness in a low frequency of anxious materialism—could be repurposed into a superefficient force for positive change. This could happen through existing media channels or through the creation of new portals on the Internet or television.

A skillful marketing campaign might package and promote the green movement, a DIY approach to problem solving, pacifism, and spiritual development as the new, hippest thing. Our revamped media would decondition great masses of people from archaic belief systems and manipulative ideologies, while educating them on subjects ranging from permaculture to metaphysics. Conflict resolution and nonviolence training will also be essential—as our current social paradigm melts down, various segments of society may be tempted to regress into violent expressions of frustration. As Thomas Hartmann noted in *The Last Hours of Ancient Sunlight*, what holds our current social system together is, ultimately, "cheap calories." If these were to become scarce due to shortages, tempers would, no doubt, become short as well. The new media paradigm must be positive as well as proactive, promoting compassion and patience and instilling great hope for a fair and equitable future, once the transition is complete. If our media machinery dramatically shifts gears to transmit a different message across the planet, this will have tremendous reverberations.

The still-unrealized social networking potential of the Internet could provide the basis for a new societal paradigm and a collaborative infrastructure, allowing for the precise orchestration of resources and humanpower in a time of crisis. Such a development would mean retooling the operational logic of the progressive movement—from struggling against the status quo to creating a new institutional framework for the imminent shift into planetary culture. Technologists have developed protocols for an invisible, underlying network that integrates user-centric profiling, data interoperability, and open standards. Such a network would allow like-minded organizations and businesses to share information at a deeper level, encourage transparency, empower individuals rather than corporations, and support the rapid development of local and virtual communities based on resource sharing, specialized interests, and other affinities.

According to this hypothesis, skillful repurposing of advanced technologies will play an essential part in the transitional process, but archaic practices have an equally central role. We must reckon with a postmodern "flatland"—to use Ken Wilber's term—where "natural hierarchies" based on the acknowledgment of different, but complementary, skill sets have collapsed. To address this complex issue, we can adapt elements of the ritual techniques and decision-making tools of indigenous tribal and aboriginal cultures and bring them into postmodern institutions and boardrooms. The design principles and organizational wisdom that allowed indigenous communities to thrive in a sustainable relationship with the natural world for thousands or tens of thousands of years constitutes an extraordinary bequest to us moderns, who have lost proper relatedness to the natural world and one another and who face the challenge of remaking those broken connections. No matter our intentions, we will not be able to accomplish this without guidance from past and present wisdom-keepers.

Since our current transition involves the recovery of the shamanic and psychic aspects of being, we may institute programs—global ceremonies—to utilize the latent capabilities of the mind for undoing the damage unleashed by industrial society. In this area as well, the living knowledge of tribal societies could prove invaluable. In the same way that we once learned to make use of electricity, we will master the subtle domains of psychic energy and use this power to transform the Earth. From indigenous cultures, we can learn the principles of utilizing ritual and trance to concentrate psychic energy, interacting with elemental forces to influence climactic conditions and alleviate suffering. The encroaching ecological crisis may compel humanity to make a quantum leap into a psychic way of being, combining scientific experimentation and shamanic participation in a new form of engagement.

Going deeper into the transition, we might expect our reality to ripple into increasingly unfamiliar forms, crossing new thresholds of novelty. As the "reality of the psyche" becomes more tangible, the development of future technologies may require the input of consciousness at a deeper level of intentionality. The proliferation of ancient and new alternative healing modalities based on the use of subtle energies (*chi*, *prana*, etheric currents, and so on) may point toward a different realization of consciousness that will have technical applications in the near future. The "Eschaton" or "singularity" theorized by Terence McKenna may be that meeting of mind, quantum mechanics, and material machinery that awaits us just ahead. Our capacity to realize such potential may depend upon the level of mental discipline and intensified psychic acuity we develop during this interim period—as Nietzsche put it, "The deed creates the doer, almost as an afterthought."

In the meantime, we are best served by adapting a tantric approach to planetary transformation—one that is collaborative instead of confrontational, absorptive rather than oppositional. The open hand, offering friendship and peaceful reconciliation, replaces the raised fist, the symbol of 1960s-style activism.

One way to envision the shift between world ages is as a snake shedding its skin: while the new scales form underneath, the surface layer has to remain functional or the snake will not survive. While the old skin of our civilization and its accompanying mind-set appear to be fraying, the system must remain in place long enough for the new texture of consciousness and its matching infrastructure to mesh together. According to this thesis, as the universal consciousness guides the process, we should have no doubt that we will succeed.

Jump Time Is Now

JEAN HOUSTON, PH.D.

According to Jean Houston, the years surrounding 2012 are what she calls "Jump Time," and we are transitioning into a new territory, a new evolution of our race and of our planet.
In the following adaptation from her book Jump Time, *she tells us how to recognize these changes in ourselves and what we can expect within our own lives and the future. As Houston says of our potential and the timing of this change, "Bursting the pods of our containment, we are ready to enter into creative partnership with the Universe and to populate our particular corner of space-time with our unique vision and capacity." She foresees radical transformation and accelerated change in the coming years. Houston shares with us her vision for the future and practical ways to identify with this Jump Time in our everyday lives.*

It may be that some of you have opened this book because you are haunted by a specter, the grand finale of the world as we have known it. You know yourselves to be people of the parenthesis, living at the end of one era but not quite at the beginning of the new one. You may have labored in your various fields to make things better, and you may have tried to understand the change of which you are a part. And yet I suspect you to be frustrated, baffled even, by what must seem the most implausible, improbable series of happenings ever. There are those who identify the expectations for 2012 with Jump Time, seeing the radical shifts happening now as pre-figural events of a still greater shift.

The maps no longer fit the territories. The only expected is the unexpected. Everything that was, isn't anymore, and everything that isn't, is coming to be. Ours is an era of quantum change, the most radical deconstruction and reconstruction the world has seen. More and more history is happening faster and faster—faster than we can make sense of it. Life paths that have contained and sustained us across the millennia are vanishing as we speak, like Gaia's species that are hourly becoming extinct. We are guests at a wake for a way of being that has been ours for hundreds, even thousands, of years.

At the same time, we know that we are the ones who must go on. It is time for us to consider what is and what may be. Our agenda is nothing less than the future. Our challenge is to cultivate the vision and lay out the practical steps necessary to move through the opening times that follow upon closing times.

Unlike many others, you may be among those who refuse to believe that chaos leads to chaos, breakdown to catastrophe. You know that you have the power to direct the process along lines different from those that the prophets of gloom proclaim as inevitable. You know that the new millennium we have entered is the intersection between worlds, between species, between ourselves and forever. You know yourself to be its pilgrims and its parents. No old formulas or stopgap solutions will suit. For a new world to be born, we must bring a new mind to bear. Nothing less will do in Jump Time.

In a Jump Time like the present, we as a species stand at a crossroads faced with radical choices, any of which promise to make tomorrow look nothing like yesterday. Personal Jump Times, moments when our life path reaches a fork and everything afterward changes, are more familiar to us. How do we react? Do we sit at the end of the known and refuse to budge? Do we walk back up the path we have been following, hoping to return to familiar territory? Or do we follow the old show-business adage that whatever happens, the show must go on?

How often in these days do we feel that our lives are like that old dream in which we are on stage in a play, dressed in a period costume, with no idea what our lines are supposed to be? Those arenas of our lives that used to hold solid expectations—professions, relationships, religions—have become capricious, the old verities lost, the anticipated outcome vanished or transformed into a mockery of itself. Our very identities seem to be shape-shifting. Standing on the stage of your old certitudes, many of you, perhaps, have experienced discomfiting surprises:

- The job or profession for which you trained and which you expected to follow for many years suddenly no longer exists.
- You fall in love with an improbable person or idea and, in an eyeblink, everything you knew or believed is gone. In the light of this tremendous circumstance, who you were yesterday bears little resemblance to who you are becoming.
- You wake up in the morning consumed by an urge to get on with it. What "it" is, you do not know, but it is barking at your heels like the hound of heaven. Something unknown is calling you, and you know you will cross continents, oceans, realities, even, to discover it.
- You are seized by concern about the disparities of the time, the widening gap between rich and poor, private affluence and public squalor. Moreover, you can't shake the conviction that you personally must do something to redress these wrongs.

- Words leap off the pages of books. Synchronicities abound. The universe is trying to tell you something, and you can no longer ignore its message.
- Your traditional religion no longer serves you, but one from an utterly foreign culture is speaking deeply to you.

Clearly you have come to the end of the road. An unknown future beckons. Way leads on to way, and only in retrospect do the turnings and returnings make any sense. I have found these things in my own life. I experienced one example of these sharp shifts several years back when I ended up on the front page of most of the newspapers in the world as the so-called "guru" and trance medium for Hillary Clinton. To set the record straight, I did not download Eleanor Roosevelt from the cosmos in the solarium of the White House. I just helped Hillary conceptualize and then edit her book *It Takes a Village to Raise a Child.* The media, however, needed a more lucrative story, so they made up a character and a spooky plot and put my name to it.

Nevertheless, the huge drama that is now unfolding on the world stage makes anyone's particular humblings look like so much hummus. The urgency behind my work, my sense of the importance of present history and its fierce momentum, helps me rise out of the flatland of my various levelings, croakings, and pratfalls and get on with it. I hope to persuade you to do the same—to stride boldly forward, even though where you might be heading and what you'll find there is not yet clear.

Perhaps you'll find, as I have, that life's jumps move you in unexpected directions. Instead of the theater of the performing arts, as I had dreamed of when I was a girl, my work has taken me into the theater of the mind, the drama of the soul, and the Broadway of cultures as they recover and share their once and future genius. In conducting basic research into human capacities and applying its findings in human development programs over the past 40 years, I have seen that lost domains of consciousness can be recovered and put to practical use. When mind and body are reloomed, psychophysical abilities for healing and self-healing, for mastering an artistic or athletic skill, and for rapid learning are quickened. I have seen that it is possible to think in images as well as words, to learn with one's whole body, and to harvest riches of creativity from interior landscapes. I have seen people discover the secrets of time—time in the body and time out of mind, in which one experiences subjectively a very short period of time as being much longer. In that short time, the mind can select, synthesize, and create in minutes what might normally take months and gain the same benefit from rehearsing an activity as if one had practiced for hours or weeks. This is what you can expect in Jump Time.

I have helped people play on the spectrum of consciousness so as not to be stuck in a single bandwidth and gain instead the option of selecting states of consciousness most conducive to concentration, mindfulness, optimal physical performance, creativity, or spiritual exercise. I have marveled that the world within is as vast as the world without and as filled with story, myth, symbol, archetype, god, and goddess. These adventures of the soul convince me that we humans have the innate equipment to be actors, directors, playwrights, and producers in the theater of the world that is Jump Time.

Beyond the laboratory and the seminar stage, part of my calling has been to study, collect, and apply human capacities as they have developed in cultures around the world. Years ago, the anthropologist Margaret Mead sent me out with letters of introduction to the elders of various cultures so that I might bring back some of the unique ways of being that ancient and indigenous societies have developed. From this beginning, my work for international development agencies gave me the chance to study firsthand how Africans walk and think and celebrate spirit; how the Chinese teach and study and paint; how Eskimos experience vivid three-dimensional inner imagery; how the Balinese learn to perform any manner of artistic endeavor so rapidly and with such high craft; how a tribe along the Amazon raises emotionally healthy children; why certain children in India raised amid traditional music develop extraordinary skill in mathematics. My work has taken me to many lands, where I treat cultures like persons and help them deepen and recover their genius, while they link with other cultures, including the emerging planetary civilization. And I treat people like cultures, helping them discover within themselves the rich strata of untapped potentials of mind and body.

No longer are potentials limited by place and culture. In Jump Time's developing hybrid world, capacities once nurtured in separate societies are available to the entire family of humankind. This is a stupendous happening, as important as the discovery of new continents during the time of the great sea journeys. For the first time in human history, the genius of the human race is available for all to harvest. These rediscovered capacities may be evolutionary accelerators, now being gathered from many places, times, and cultures to awaken our species to who we are and what we yet may be and do. Often, however, it is not comfortable. We can, for a time, find ourselves to be strangers in a very strange land, wishing we could return to the comforts of a more insular and familiar worldview. And yet, when we get beyond the shutterings of our local cultural trance, we gain the courage to nurture the emerging forms of the possible human and the possible society.

With the present move toward planetization and the entire spectrum of historical social development at hand, I believe that the world is set for the radical transformation that I call

Jump Time. In this time of accelerating change, all cultures, regardless of their social or economic level, have something of supreme value to offer the whole. I have observed too many members of European-derived cultures who revel in technique and objective mastery but sadly lack in the spiritual awareness and subjective complexity found in aboriginal cultures belonging to earlier stages of historical development. True, the tragic intervention of colonialism dimmed these cultures for a while, but more recently, Maoris, Australian aborigines, and Native Americans, among others, are restoring their cultural wisdom. As the genius of many cultures is brought together, as I believe is now happening, the current crisis of social breakdown and moral disorder can be transformed into the creative symbiosis of the coming world civilization, cultures seeding each other, while preserving and enhancing their individual style and differences.

So, what is Jump Time? It is the changing of the guard on every level, in which every "given" is quite literally up for grabs. It is the momentum behind the drama of the world, the breakdown and breakthrough of every old way of being, knowing, relating, governing, and believing. It shakes the foundations of all and everything. And it allows for another order of reality to come into time.

Jump Time is a whole system transition, a condition of interactive change that affects every aspect of life as we know it. The vision of change I am describing is generally optimistic. It focuses on the emergence of patterns of possibility never before available to the Earth's people as a whole. However, this optimism is based on the hard-nosed recognition that virtually every known institution and way of being is currently in a state of deconstruction and breakdown. And yet, given the scientific, technological, cross-cultural, and social tools at hand,

and given, too, that humanity is searching as never before to cooperate in so many areas, it seems feasible to me that we may be ready to integrate inner and outer dimensions of life in ways that infuse new depth into psychological and spiritual growth and new purpose and responsibility into social transformation.

As I see it, our current stage of collective growth is being propelled by five forces, which, taken together, determine the direction of our jump into the future and where we may land:

1. *The Evolutionary Pulse from Earth and Universe.* Basic to Jump Time is the belief that we humans are not alone as we face the massive transition that is upon us. Rather, we are embedded in a larger ecology of being, its motive force arising simultaneously from the planet, which is our birthplace, and the stars, which are our destination. Pulsed by Earth and Universe toward a new stage of growth, we are waking up to the realization that we can become partners in creation— stewards of the Earth's well-being and conscious participants in the cosmic epic of evolution. As ancient peoples have always known, the story is bigger than all of us, and yet desires our engagement, our love, and our commitment. People everywhere are feeling a call, a quickening, an energy that brooks no whiny naysaying, an invitation to the evolutionary impulse that drives our growth and suggests how we might learn to ride and direct its energy.

2. *The Repatterning of Human Nature.* The second force of Jump Time pushes us to discover and utilize our dormant or little-used capacities and to come to a more comprehensive understanding of our place and responsibilities in this world and time. In Jump Time, capacities that belonged to the few must become the

province and requirement of the many if we are to survive the next hundred years. We must each learn to tap into the creative workshops of the mind to solve problems and to bring forth art, poetry, invention. We must discover ways to feel at home with anyone, anywhere, at any time. Most people, given opportunity and training, can learn to think, feel, and know in new ways; function in their bodies with better use and awareness; become more creative, more imaginative, and aspire within realistic limits to a much larger awareness, one that is better equipped to deal with the complex challenges of life. The consciousness that solves a problem cannot be the same consciousness that created it. But the consciousness that can rise to this occasion needs models of its own matured possibilities, visions of what humans can be and do who go beyond the limitations of academic excellence or dogged persistence to attain certain goals.

3. *The Regenesis of Society*. As the self is repatterned, the ways we relate to each other are necessarily shifting as well, toward the discovery of new styles of interpersonal connection and new ways of being in community given a global society. The movement seems to be from the egocentric and the ethnocentric to the worldcentric—a fundamental change in the nature of civilization, compelling a passage beyond the mind-set and institutions of millennia. Critical to this reformation is a true partnership society, in which women join men in the full social agenda. Since women tend to emphasize process over product, being rather than doing, deepening rather than end-goaling, it is inevitable that as a result of this partnership, linear,

sequential solutions will yield to the knowing that comes from seeing things in whole constellations rather than as discrete facts. The consciousness engendered by this comprehensive vision is well adapted to orchestrating the multiple variables and getting along with the multicultural realities of the modern world. It raises hope for forgiveness between and healing among nations and ethnic groups. Essential to this matured consciousness is moral and ethical growth toward an empathy between individuals and nations that honors the golden rule of human interchange. The regenesis of social forms also asks that governments no longer engage in social engineering to fix specific problems, but rather that they understand the world as an ecology, a complex adaptive system in which global awareness is applied to local concerns. Here we need models of a new order of relationships and their place in a possible society, one in which male and female, science and spirituality, economics and ecology, civic participation and personal growth come together in an integral and interdependent matrix for the benefit of all.

4. *The Breakdown of the Membrane.* In the interdependent world of Jump Time, old barriers are dissolving, along with the phobias that sustained them. The new technologies of instant communication and exchange of information enable people to join minds and hearts in mutual discovery and creation. In the "Internetted" world, more and more people are coming to accept the benefits of cultural diversity and to adopt a more inclusive worldview, inspiring human nature with renewed hope and caring. What began in migrations and global economics is fast

becoming a worldwide network of individuals and institutions, quickened by the desire to create a new social paradigm in which humanity and the Earth are each enhanced within the context of a collective destiny. As the membrane of old forms breaks down, a more complex and inclusive global organism comes into being. As living cells within this new organism, we are rescaled to Earth-wide proportions in our responsiveness—and our responsibilities.

5. *The Breakthrough of the Depths.* The depths are breaking through most apparently in the spiritual renaissance that is occurring everywhere in Jump Time. Not since the days of Plato, Buddha, and Confucius, some 2,500 years ago, has there been such an uprising of spiritual yearning. And what with the inevitable cross-fertilization of the wisdom and practices of world spiritual traditions, more and more people are gaining access to the source of our being and becoming. As a new story begins, the mythopoetic understandings of many cultures rise and converge. Archetypal ideas and symbols spring into consciousness or are consciously sought in the popular culture. Either the depths are rising because the time requires it, or we are being impelled by the times to explore those depths. However it is occurring, our spiritual preoccupations are essential, given that we need a spiritual renaissance commensurate with our technological developments to give us a sufficiency of inner inspiration to guide our expanding outer forms.

The world has known other Jump Times, but never so consciously or with so much to gain or to lose. The Earth is a hothouse now. Six billion members of the human family and rising, congregated together on a spinning ball, in stress, in ferment, caught between what was and what is yet to be. It is time to ask the great questions: How can we make a better world? What must we do to serve the larger story? These questions help us clarify and define. They prompt us to articulate goals lofty enough to lift us out of petty preoccupations and unite us in pursuit of objectives worthy of our best efforts.

The world is hungry for vision. At a time when whole systems are in transition and global forces challenge all authority, there is an insistence in the mud, contractions shiver through the Earth womb, patterns of possibility strain to emerge from the rough clay of changing social structures.

It is a matter of *kairos*, the potent time for fortuitous happenings. In ancient Greek, kairos referred to that moment when the shuttle passes through the openings in the warp and woof threads. In the loaded time when such things happen, the new fabric can take form.

Right now is the right time to make things right. I invite you to join me on the stage of forever, where what we do may raise the curtain on the world's next act.

A Vision for Humanity

BARBARA MARX HUBBARD

Can we be conscious participants in our own evolution?
Leading futurist, teacher, and author Barbara Marx
Hubbard, the president of the Foundation for Conscious
Evolution and a former nominee for the vice presidency
of the United States, speaks to the evolutionary
possibilities of the coming change of 2012. In the
following essay, she addresses the evolution of consciousness,
society, the "egoic mutants," and the human race. She
also tells us what we must do specifically to prepare
for the new species we face . . . or might become.

Humanity is facing unprecedented, evolutionary changes. It is amazing—out of the famous Mayan prophecy has come the indication that we are facing the end of this world as we know it and the beginning of the new world by 2012. This date corresponds to environmental and social predictions of breakdowns and breakthroughs now widely known. Let's assume there is some validity in these prophecies. What vision of the future, of the new world, might we see so that we can place our attention upon this vision as a strange attractor to carry us through this critical transition?

I see a vision of the future that springs from witnessing the awesome mystery of our past—from 14 billion years of astonishing evolution, and from the genesis and unfolding of the cosmos to the present and beyond. The stunning realization is that from "no thing at all" has come everything that has been, is now, and will be. The force, intelligence, or process that produced this universe is still operative in us. We are the universe in person. We are manifestations of that mystery. We are the cosmos becoming self-reflective and attempting to understand its own origin, birth, and direction and how it can best participate consciously in the process of creation.

We are receiving guidelines for our own conscious participation in evolution. We learn from our past. How did nature evolve from subatomic particles to us? When we comprehend this new story of creation, from the first flaring forth to us now, we see a recurring pattern of ever-greater complexity, bringing forth ever-higher consciousness and freedom. From atom to molecule to cell to animal to human, to the Buddha and Christ and so many others, to us now, we are at the threshold of devolution or transformation. "Evolution's arrow," as John Stewart calls it, has manifested a progression toward more comprehensive whole systems through ever-greater cooperation and synergy, quantum jump by quantum jump. Herein lies the foundation of our hope for an immeasurable, ever-evolving future.

We also notice that crises precede transformation. Problems are evolutionary drivers. Intensifying pressure leads either to devolution and extinction or evolution and transformation. This is one reason that 2012 might be earthshaking and shocking to some. Furthermore, we see that evolution creates radical newness. Life out of prelife, human life out of animal life. Is it possible that from our species, a new human and a new humanity are now emerging? We can see the sequence from Homo habilis, Homo erectus, Homo neanderthal, Homo sapiens, to Homo sapiens sapiens. Why should it stop with us? Is it possible that a new species has been gestating in the womb of self-consciousness through the great avatars, mystics, and visionaries for thousands of years, and that now, due to the crisis, this new human is being born in millions of us? It has been predicted as Homo progressivus (Teilhard de Chardin), Gnostic human (Sri Aurobindo), Homo sapiens sapiens sapiens (Peter Russell), Homo noeticus (John White), Homo universalis (my own favorite name). It is possible that the basic shift point in this prophecy is the emergence of the "universal human," by whatever name it is called, as a "new norm." Perhaps we will reach critical mass for this human by 2012.

The fact is we see ourselves coming to the end of the viability and even survivability of ourselves in our current stage of self-centered self-consciousness. We are becoming "egoic mutants," as Alan Lithman calls current leadership, gaining powers we used to attribute to gods while we remain in the same state of consciousness as our ancestors.

If we imagine ourselves only a few years in the future, even by 2012, providing we do not destroy ourselves first, we see the outlines of a vastly enhanced humanity, applying the growing edge of human consciousness, capacities, and evolutionary technologies to life-affirming goals, resulting in a quantum transformation of the human species itself.

All the major spiritual traditions originated before we had any notion of the evolutionary process, of the billions of galaxies, of the fact that Earth is not the center of the universe, but a small planet revolving around a sun in the midst of a multidimensional universe of awesome mystery. They also all originated before we gained the power to cause our own extinction or transformation, before the advent of *conscious* evolution, as we are now being called upon to understand how to sustain and fulfill the potential of life on Earth. Something new is happening.

The first element of my vision of humanity is that Homo sapiens is giving birth within many of us to a more "universal human," a human who will eventually be as different from us as we are from the Neanderthal.

This universal human comes from every race, nation, religion, and discipline. Some of the qualities of its being are now becoming apparent. A universal human feels connected through the heart to the whole of life and is awakened from within by a passionate and loving desire to express unique creativity for the good of the self and the whole. For this emerging human, the veil between dimensions of reality is thinning. We are becoming multidimensional beings, having expanded reality and peak experiences, receiving higher guidance and epiphanies vital to guide evolution now. The universal human is developing evolutionary consciousness and evolutionary spirituality that will give us a cosmic whole organism, scientifically and spiritually based in awareness. In addition, as women have fewer children and live longer lives, we are shifting from maximum procreation toward cocreation, from self-reproduction toward self-evolution through life purpose and creative self-expression, giving birth to the authentic feminine self and her expression in the world. Men, in their own way, are experiencing this same inner evolution. We are all participating in what Carl Jung called the continuing incarnation of the Self. The Self we projected outward upon the gods is coming home as our own developmental potential.

THE EMERGING SYNERGY OF PLANETARY EVOLUTION

Let's place this emerging human in the context of planetary evolution of 2012. We can see that every system is breaking down—health care, environmental, economic, governmental, and so forth. We also notice, if we look carefully, that innovations and creative solutions are arising in every field. Although disconnected, underfunded, and undercommunicated, they are nonetheless *here* now and do function.

We learn from Ilya Prigogine, the Nobel prize–winning chemist, that when systems become more dysfunctional, they tend to use ever more of their energy in handling their own dysfunctionalities. At the same time, mutations or social innovations crop up throughout the system. Nature has a tendency to converge upon that which is creative, to manifest new whole systems at a higher degree of order. We are not very far, right now, from the possibility of the nonlinear, exponential interaction of innovating elements, leading to the transformation of our social systems. The very crises we are facing are causing innovations and breakthrough solutions, which are even now networking and connecting. With just one more degree of the networking of the networks, we are approaching a breakthrough in "co-intelligence," in the collective capacity of that which is emergent, compassionate, and creative.

In 1984, when I ran for and was selected as a nominee for vice president of the United States on the Democratic ticket, I proposed a new social function: a "Peace Room," which was eventually to become as sophisticated as our war rooms. Its purpose is to scan for, map, connect, and communicate what is working to heal and evolve our world. We can imagine that the nonlinear, exponential interaction of innovations that work could speed up time and cause an apparent sudden shift at precisely the time frame of 2012.

Now that the Internet is functioning, it is possible to do just this. We may be at the threshold of a jump in social synergy—of the coming together of separate groups and functions to form an emerging whole organic system. We know that the whole is greater than, different from, and unpredictable from the sum of its parts. We can expect radical newness.

We can move toward a more synergistic democracy, a democracy based on innovations that are emergent in every field, a democracy the purpose of which is to encourage everyone to do and be their best within the evolving whole. Imagine a cocreative society that emancipates the greatest untapped resource on Earth: the suppressed or misdirected capacity and creativity of the peoples of the world. Imagine each person being free to do and be his or her best. Nature takes jumps through greater synergy, atoms join together to make molecules, which join together to make cells, which make animals, and now we are being connected as one living system on planet Earth. Synergy creates newness. Once there was no life, then there was life; once there were no humans, then we appeared. Evolution teaches us to expect the unexpected, to anticipate the new.

Beyond immediate innovations, there are some genuinely radical breakthroughs at the horizon. As we run out of fossil fuels and develop renewable sources of energy, we are at the threshold of a quantum jump. It may be possible to tap into zero-point energy, the infinite energy available in the "plenum" of space. We may be able to extend our life span, to live and settle our solar system, and to move into the galaxy. We may be able to reach out and contact other life, other dimensions of reality. Quite literally, Earth is giving birth to a universal humanity, a species capable of coevolving with nature and cocreating with spirit—or of devolving and self-destructing.

One of the fundamental keys to all of this is the evolution of the person; the appearance of greater numbers of humans capable of comprehensive compassion and consciousness to guide these new powers and to develop new social systems and structures to enhance Earth life. If we project ourselves forward as universal humans, pioneers in the building of a new culture and civilization—a universal humanity—both on this Earth and eventually in the universe beyond our home planet, we can see the outlines of a quantum transformation.

A long-range vision for humanity sees us stabilizing in cosmic universal consciousness. The inner impulse of evolution is becoming self-aware in us. We are gaining access to the implicit order or creative intelligence of the universe, and it is manifesting in our own guidance and actions.

We are an Earth/Space, and eventually a universal, species. As we restore the Earth and free the creativity of our people, we realize that we are literally like the fish coming out of the sea onto the dry land. We will build new worlds on this Earth and in space; we will extend the Earth environment into the solar system and beyond. We will cocreate with nature a humanity that lives beyond our planet. As the Sun eventually expands and destroys all Earth life and all solar systems, we will be a galactic species.

Our combined intelligence, extended through our computers, will be a quantum jump beyond what we have ever known.

We will experience cosmic contact and find that we are not alone in the universe. The trillions of star systems may actually be mothering wombs for new life on countless planets. Maybe we are too young to have yet encountered others. When a baby is just born, its nervous system is not yet mature enough to see "other life." But out of hunger and the need to coordinate, its nervous system links up, the infant opens its eyes, sees its mother, and smiles. The infant recognizes what it has never seen before. We may be on the threshold of our first "planetary smile," a first planetary awakening.

It is my intuition that as we mature as a universal species, we will learn resonance and mass connectivity here on Earth. Through a more harmonious order, we will be in touch with whatever other life there may be in the universe. Instead of contact with alien cultures, I foresee us attracting higher life that has dimensions of our own higher state of being. If we do not mature ethically, even with all our high technology, we won't make it through the next stage of our evolution. If we do learn ethical conscious evolution, our morality and compassion will have evolved beyond the current warring humanity. This may be the viability test for our birth into universal life. If and when we learn to love one another, to resonate with the higher aspects of our own being, my sense is that we will find that we are in a universe full of life, that we are reproducing and evolving Earth life in a universe of fabulous diversity, and that our species itself will diversify and multiply like star seeds throughout the universe.

Beyond that, I can see that "God is reproducing godlings." We are to become cocreators with the process of creation until we have joined together with others doing the same in a universe that is infinite and self-creating through all of us— worlds without end.

HOW DO WE GET READY FOR 2012?
Realize that the higher self, the inner voice, is you. Bring your projections home as your own developmental potential. Ask that deeper self to take dominion within your personality selves, and shift your identity from your ego to your essence. This is the key step to being born as a more universal human. Then, find life purpose. Ask to know what you are born to do. Follow the compass of joy. Seek out those who affirm the highest in you. Where two or more are gathered in the name of that

highest dimension or your being there, "I Am" is in your midst. Everyone is needed. Everyone is called. The greatest blessing any one of us can have is to say "yes" to the dormant potential within us. Reach out to those who attract you, and connect with as many others doing the same as you possibly can.

2012 Awakening to Greater Reality

MEG BLACKBURN LOSEY, PH.D.

With evolution linked to some 2012 prophecies, many wonder just what parts of our species will evolve. Meg Blackburn Losey sides with the popular belief that it will be our consciousness. The key, she says, is in our DNA. How do we assist this change or identify it within ourselves? Dr. Losey gives us practical, wise, everyday ways to detect this power of consciousness and bring it to light in our own lives.

What if one day we awoke and found that our perceptions of reality had changed? Not subtle little changes, but new and different ways of seeing not only the third-dimensional world, but beyond? What if we began to see shadows of other people, or holographic beings; heard voices, either all in a jumble and just out of earshot, as if at a party, or talking to us directly, telling us about time and space, the formation of reality, the possibilities of conscious travel in time and space?

What if, suddenly, we just *knew* things, such as what was going to happen next, what others were thinking, or even future events? What if we began to see or feel subtle energies, became aware of even the tiniest changes inside of our bodies and around us? Would we be certifiably insane? Absolutely and positively not.

Every day, more and more people experience varying extremes of these new and strange perceptions. What is going on? Why would this happen? What does it mean? Are we ready for such drastic changes within our scope of reality? Why is this happening to some people and not to others? Is there something that we can or should do about it? Are we falling victim to some strange sense of consciousness that is out of control? No way.

There is a shift of consciousness in occurrence. Many people believe that the year 2012 will be the end of the world, signifying mass destruction and major changes on our planet. While there may be certain events involving Earth changes, our planet has been changing her complexion since the dawn of creation. The fact that we are here doesn't change that.

The true meaning of the ancient prophecies relating to the year 2012 is that we are going to take a major leap in human evolution. As we enter into a shift of ages, we have the opportunity, as a collective, to discontinue ignorant existence. As this shift takes place, we will find that how we live in our world and how we perceive our world are going to change drastically from our previous and current viewpoints.

A change of ages requires a change of consciousness—a different way of existing altogether. First let us explore some simple basics.

Let's face it, our world is at a pivotal point between survival of the human race or its destruction, by our own hand. Some of this is due to political decisions made to fatten the coffers of the few; but more of the human dilemma is due to the complacency of humanity. In many ways, we have become comfortable with our many conveniences. We let "them" take care of everything. "They" are screwing up, *not us*; and "they" can certainly fix that. We have created an illusion of safety and security that no longer works.

As we move into the shift of ages, up to and beyond 2012, there are amazing happenings within us that will carry us to new and different forms of consciousness.

WE ALREADY HAVE A BOUNDLESS CONSCIOUSNESS

Our consciousness is separate from our thought process. Our consciousness knows no boundaries and is *always* aware of what is happening in our world and beyond, in other dimensional reality. In fact, our consciousness exists fully within all levels of reality, all of the time; and it is doing its best to get us to listen to the clues and signals that it picks up in all of creation.

What that means is that 24-7, our consciousness acts as our supermonitor to let us know when we need to alter our path or our way of perceiving any given moment. Because our evolutionary processes were originally based upon our survival, our consciousness has been forced into the background. Since our consciousness remained hidden from our everyday awareness, previous generations were not able to hear the messages that our consciousness continually delivers to us.

Early in our existence, we were beings who communicated purely with telepathic abilities. We even transferred data via subtle energy that was carried to us by external electromagnetic fields. As that energy passed through us, it literally became part of us, or at least deposited messages within us as the energy flowed through our bodies. Our awareness had no limits, because we were extrasensory existent—we were metaexistant. We subsisted outside of measurable bounds. Even now, when we dream and when we have extrasensory experiences, our consciousness takes us to realms far beyond this *now*. That is because, in these situations, we have let our defensive guard down, our brain is out of the way, and our consciousness is able to communicate with us more fully.

Over time, we became more and more dense. We manifested as physical beings, solid forms that were vulnerable within our environments. We had to learn to protect ourselves from predators and an unstable planet.

Later, as many of us began to migrate, the root races from which we descended began to move from their places of birth and out into the world, often meeting others along the way, learning to trade and exchange resources, developing commerce and, with that, communication through language.

As we learned to communicate this way, the art of subtlety was born. Because of language issues, we wondered what people who were not of our immediate clan really meant, if they were being fair with us, or if there was more we didn't know or wouldn't get in our trading.

Conversely, we worried if we were giving too much away, so we began to think about our communications, even plan them. Linear thought became the norm. We began to plan into the future or worry about the past. Our egos stepped in and began to cloud things even further—we forgot the abilities and wisdom of our consciousness.

Our egos have lied to us for millennia, judging and misjudging every moment of our waking existences to the point where many of us have trouble making decisions. We have lost the ability to be fully present in any given moment because we have become filled with the chaos of our egos and thinking minds. That chaos reprogrammed our energy systems until we began functioning in a less-intuitive way. We became desensitized to ourselves and our environments, missing the subtleties around us that used to guide us exquisitely. The lack of sensitivity that has become the norm often leaves us feeling out of balance.

Due to our evolutionary development and desensitization, parts of our brains were less needed and became less active; ultimately shutting down more and more interrelated activity until, eventually, we reached stages of evolution when we used only a small percentage of our brain matter. We lost awareness of our innate extrasensory abilities and began to rely only upon our physical senses.

THE KEY IS IN OUR DNA

Like our brains, our DNA—the genetically coded material within us that commands how we look, how we act, and a multitude of other parts of our earthly experience—is also a very important part of our universal communication system. Plain and simple, our DNA responds both to our consciousness and our thinking self.

Our consciousness always knows it is perfect, while our minds are constantly telling us differently. Our minds are the thinking, logical parts of ourselves, the parts of us that, combined with our egos, tell us that we are safe and assist us to understand what we think we are doing. Our thinking self is limited to our brain matter and subject to data that is received during our life experiences. Our consciousness is not limited to physical experience and operates within the overall universal construct. It is unlimited in nature and aware of greater reality all of the time. Our consciousness is basically constructed of light energy, which works on principals of electromagnetics. When we are happy in our lives, we feel well and whole. Yet often, if we develop belief systems that set us aside from all others or if we feel we are less than others in some way, we become miserable in our experience, and communications throughout our bodies reflect our misery through sickness, weight gain, or a general not working very well. Often, how we present ourselves to the world is a direct reflection of our inner selves.

When our brains are running the show, the doors to our true being, our consciousness, are closed. We lose access to higher knowing and remembering our true being. By virtue of its evolution, humanity in general has closed its doors to greater reality.

Because our DNA is responsive to our experiences similarly to our brains, as the etheric wiring inside of our brains was used less and less and parts of our brains began to go dormant, so did parts of our DNA.

Slowly, communications along our DNA went from leaping actively from one protein segment of the ladder to another in holographic form to a more linear form of communication. Straight-line communication developed from one protein segment of the DNA ladder to the next, not only limiting communications on a biophysical level, but also limiting our conscious awareness. Instead of having omnipotent awareness, we became limited to our life experience. For instance, when we have an interaction with another person, our entire body is sensing and even listening energetically. Our DNA responds from our human perceptions. When we feel in danger, adrenaline pumps through our bodies. This is a physical reaction and nothing more. Our consciousness knows it is not in danger and would respond with information about what the other person really means, really wants, and, instead of reacting as if we were in danger, we would instantly know how to deal with the situation—or if we needed to at all.

In turn, our RNA—the part of our genetic system that serves as the communication network throughout our bodies and ultimately directly to the DNA—changed its communicative pattern as well.

The inherent electromagnetic energy fields that existed in our DNA strands became extremely weak and, in some cases, nonexistent. Like our brains, the sections of DNA we don't use remained intact but inactive. Because of this, we became both mentally and genetically limited, using less than 10 percent of our DNA materials.

That is, until now.

As we journey toward the year 2012, there are many factors that contribute to changes in our electrical, electromagnetic, and genetic makeup and the inherent relationships among each of them. Innate within each of us, and particularly within our DNA, exist ancient memories of our former selves. These memories cause us to constantly seek who and what we are. Often, we extrapolate our divine selves into the form of an external god or divine something. But, *we* are what we seek.

As we evolve in a circular format, moving closer and closer to our beginnings—our perfect selves—the energy of all nature continues to operate in a spiral formation as it has since the beginning of time. We evolve closer and closer to who we were in the beginning. In fact, before the beginning.

When we began as spirit, as light, as pure consciousness, we were unencumbered by physical bodies or limitations of any kind. We were the epitome of perfection. Subtle energies have always moved through us on a continual basis. As those energies traverse us, they carry messages to our entire being. As we get closer to 2012, the messages are becoming stronger and clearer. They are telling us to awaken, as if we have been asleep for many millennia. And many, many of us are hearing the clarion call to greater reality.

HOW WILL CHANGE BEGIN?

Awakening to greater realities beyond our local reality has many variances in symptoms and abilities. Common symptoms of awakening are changes in how we see the world, changes in how our bodies respond to food and external stimuli. Often, as our consciousness shifts, our bodies have a difficult time keeping up. Because our bodies are our densest nature, they cannot adapt as quickly to energetic changes. An extreme example of what this is like is when we receive an electrical shock. The electrical energy is too much for our bodies. On a more subtle basis, electromagnetic changes actually cause stimulation of various types to our bodies. When this type of stimulation is new or different, our bodies react by sensing discomfort. Stiffness in the upper back, shoulders, and neck are quite common, as are transient joint and/or muscle pain, digestive issues, foggy-headedness, short-term memory loss, and reality glitches.

Reality glitches are best described by this example: We enter into another room to do something and find that it is already done, yet we have no memory of doing it. Conversely, we think we have done something and find that we did not do it, yet our memory tells us that we in fact did. Not only this writer has had these experiences, but many, many others have as well. As energy shifts take place, this type of occurrence seems off and on. When the veils of reality move or change in density, reality becomes questionable. Our consciousness knows what is happening, but our thinking minds begin to question our sanity.

Another example is remembering a conversation that we had with a person, only to find ourselves in the exact same conversation with the same person in the future. This time, though, the conversation has some variances from our memory of it. In truth, we had the original conversation outside of the third dimension, in the future, in another dimensional reality! Our consciousness leaped the time continuum. The same applies for other déjà-vu experiences. Our consciousness has already experienced our future before our third-dimensional selves got there.

Because our consciousness doesn't follow the same rules as most other forms of energy, it can go wherever and whenever it wishes. As we progress in our evolution, we begin to be more and more aware of our multidimensional selves. The doors are opening again, and we have begun to notice.

Other symptoms of greater reality seeping in are knowings and intuitive feelings that come from deep inside of us. We don't know how we know things; we just do.

As our circular evolution continues toward 2012 and many of us begin to respond to that clarion call, our entire beings are opening to glimpse clues of our true nature. Our brains and our DNA begin to reawaken. That includes seeing through the etheric veils and into other dimensions, spontaneous gifts of telepathy, remote viewing, full sensory knowing of things to come and of others' thoughts and motivations. The future, in fact, holds an unlimited array of new types of sensory perception. This is what I call our Seventh Sense, otherwise known as our multidimensional, universal sense. Our Seventh Sense will evolve more as we near 2012.

As we progress toward the shift in consciousness, more and more people are awakening. I have been contacted by countless of them as they have wondered about their stability, or conversely, are very excited at their peeks at other realities. Some people with whom I have worked have had their awareness open so quickly that their entire reality, in fact their lives, have changed dramatically, seemingly overnight. In my personal experience, awareness has leaped time and again, to the point that it often feels as if I am online all of the time. Some people, who have been gifted all their lives, are learning to use that giftedness in ways they had never dreamed. Others are awakening to the beginnings of their greater senses in varying extremes. Sometimes, reality changes quickly as voices are heard, beings are seen, first out of the corner of the eye, then perhaps later, more clearly. For others, knowingness seeps into human awareness, stretching everyday reality until awareness of other realities becomes commonplace.

Often, the illusions we have created in our everyday reality come crashing down, revealing untruth, self-delusion, or errant belief systems that were based upon the reality imposed by our ancestral lineage, our caregivers, clergy, and others who did the best they could with what they knew.

When our reality changes so drastically, it is a challenge to function in the third-dimensional world until we are able to find balance with our new awareness. At the same time, we find that living in the world we once knew requires an entirely new set of skills. Not only that, but our value systems change. The things we once thought were important no longer seem to serve us. For example, we are conditioned by society, by our parents, and by others to achieve goals—that success is our purpose. But no one ever told us what success really looks like. We began to feel empty, attempting to fill those needs with things, possessions, or seeking out experiences that provide more and more stimuli. Extreme sports are good examples. Having to own the biggest, the best of everything is another. With our new heightened awareness, we begin to realize that nothing is really missing. We begin to feel full and whole on the inside, and our needs become simpler and our perspectives fuller.

No longer are the illusions of third-dimensional living acceptable. There comes a time when boundaries of consciousness become irretrievably crossed. Truth in living becomes a requirement. Those things we accepted as a matter of course are no longer acceptable. In fact, for many of us, our entire perception of life is changing.

OUR ENERGY CHANGES IN 2012

As our bodies respond to the energetic changes, our electrical patterning begins to change. We create more and more electromagnetic activity just by existing. This electromagnetic activity is allowing our consciousness to become more and more accessible to us. Because of this, we are becoming more highly aware of subtle realities. At least, they seem subtle to us.

Further, as our electromagnetic fields change and our electrical patterning is altered, another amazing phenomenon occurs. Ultimately, we are composed of sets of harmonics that are a lot like intricate musical chords. Each of us is a unique combination of energy and light. Energy and light exist at different frequencies; they are different colors and, if we could hear them, different sounds.

As we evolve and the electromagnetic energy and electrical energy within us respond to the changes, our harmonic frequencies change as well. They become higher, more similar to the harmonics of our source. Our harmonic vibrations become more and more refined, and with that, our perceptions, our abilities to perceive our environments, and ourselves become more developed as well. We begin to live in our world very differently than we have at any other time in our lives.

As our vibrations rise and we begin to relate differently in our world, all that is happening within us is communicated in the form of subtle energies through the entirety of creation. And creation responds. The reality we perceive begins to change, not only within us, but also reflected around us.

When these kinds of changes occur, there is a universal response, from the tiniest aspects of creation to the greatest. First, we begin to notice little things changing. Our perceptions change. The people in our lives change. Those people who are no longer vibrating at or near our level seem to fall away, and we begin to attract others who are harmonically constructed in a more similar fashion to ourselves—who are more like us in perception.

GLOBAL SHIFT—DIMENSIONAL SHIFT

As these spontaneous changes occur and more and more people who are of similar harmonics come together and interact, a growing number of people are awakening to their true nature. And they, too, begin to gather. Ultimately, there will be a moment in eternal time when enough people have awakened that a spontaneous shift will occur in the reality in which we live. Many call this a dimensional shift; in fact, it is.

Any time there is a shift of such magnitude, it is not only dimensional but also universal. The entirety of creation changes in some way to reflect the messaging it has received, and new reality is created.

It is not this writer's contention or belief that 2012 means that the world will end or that humanity will disappear off the face of the planet. Instead, we are about to take a great leap in consciousness, one that will change reality as we know it. The possibility exists that humanity is on a path to reunion with the very one from which we all came, that we are evolving to a point when we will spontaneously shift to a complete change in perception and even existence. Why not?

The end of the Mayan calendar and those words of our ancient predecessors did not mean that we would disappear from our planet; rather, we could step into a new form of reality and the age of ignorance in which we have lived for millennia would come to an end.

In this now, as always, humanity has a choice of direction and the culmination of all of its evolution up to this moment. More important, we are at a crux point in the very beingness we so treasure. We can easily accept the changes that are occurring within us by choosing to create a new and different reality that honors each of us as part of the greater whole; or we can choose to fight our awakening, kicking and screaming all of the way to oblivion. It has always been a choice.

In order to move through the coming changes of 2012 with ease, there are eight keys to remember. They are as follows:

We must acknowledge our self-perfection. We are and always have been one with Source.

Accept the journey for which we have come. Why fight the very things that we have come to learn?

Maintain personal integrity. We must trust the truth that is ours among all other things and give ourselves an opportunity to find the truth within us.

Be that which we are, not that which we perceive others would see us. We are created of light, of grace, and of that there cannot be imperfection, only that which is of spirit. We do not need to improve ourselves, only to acknowledge that which is our God-self, our perfect being.

We must acknowledge our true value. This is different from accepting perfection. Our value is how we fit within the world inside of us as well as everything outside of us. To perceive that our value is greater than another's, or less than someone else's, brings us to lack of everything else.

Accept our power. We are great and mighty. True power comes not from ego, but from the collective spirit. True power is gentle power. We are of the light, the essence from which all things are made. To fear inner power is to suggest that we are less than all other things. In truth, power is of grace, not of abusiveness or negative use, and our grace is unconditional. True power is that which is love, intentional living, and creating.

Take our value, our perfection, our power, our grace into your world. In the now, change only comes from practice of change. What this means is that to effect change we must embody it. We must walk our talk, not hide that which we know.

Love ourselves and touch everyone we encounter with love. As all energy exchanges, what will we accept from others, and what will we leave behind? We can see all others as mirrors of ourselves, that their pain also resides somewhere within us, that their joy is in our hearts as well. This is why random acts of kindness make such a difference. How many times have we said, "There by the grace of God go I"? It is true. It has always been so.

Most of all, remember to breathe! When we encounter experiences that are new and different, we have a tendency to hold our breath. When we do this, we literally pause the flow of our internal energy system. If we hold our breath enough times, our energy begins to compact, and we lose our ability to act intuitively. Remembering to take several deep breaths, even four or five times a day, assists us in clearing unwanted densities in our field.

Perhaps if we could hear the beauty of our inner and combined harmonies, we would hear the most beautiful symphony that was ever created. The color and light would melt even the most hardened of hearts, and we would stop in our tracks and consciously join the chorus of the universe, wanting nothing more than to be immersed within the music.

And the calendar would begin again.

The Great Turning
as Compass and Lens

JOANNA R. MACY, PH.D.

*As the evolution of the human race becomes inextricably bound to
2012 in the popular consciousness, the evolution of the environment
must be addressed. Eco-philosopher Joanna Macy, famous for her
spiritually infused eco-awareness, believes we are on the verge of
a great change. As Macy writes in the following essay, she sees the
"Great Turning" as "the essential adventure of our time: the shift
from the industrial-growth society to a life-sustaining society."*

What it means to be alive at a moment of global crisis and
possibility:

> *Thinking of the Great Turning reminds me I don't have to save
> the world by myself; then there's more energy for my little piece of
> it—getting the military out of my son's school.*
>
> —Antirecruitment activist in San Francisco

> *It strengthens me to see my work for renewable energy in the
> context of e grande virada.*
>
> —Corporate consultant in Brazil

*I love telling the children in our eco-camp that their restoration
project is part of* die grosse Wandlung, *and they are part of it, too.*
—Teacher in Germany's Black Forest

Now I recognize el gran cambio *right here in Barcelona, and at
the same time it links me with activists around the world. I feel
less isolated.*
—Community organizer in Spain

E grande virada, die grosse Wandlung, el gran cambio . . .
Wherever I go, in every group I work with, the Great Turning
becomes more rewarding as a conceptual frame. It is a name
for the transition from the industrial-growth society to a life-
sustaining society. It identifies the shift from a self-destroying
political economy to one in harmony with the Earth and enduring for the future. It unites and includes all the actions being
taken to honor and preserve life on Earth. It is the essential
adventure of our time.

Of course, most people involved in this adventure do not
call it the Great Turning. They do not need that name in order
to fight for survival and to fashion the forms of a sane and
decent future. Yet, more and more of us are finding that concept to be wonderfully useful. As teacher, activist, and mother,
the Great Turning helps me see what the physical eye cannot:
the larger forces at play and the direction they are taking. At
the same time, it sharpens my perception of the actual concrete
ways people are engaging in this global shift. In other words, it
serves me as both compass and lens.

THE BIG PICTURE
From the countless social and environmental issues that compete for attention, we can take on isolated causes and fight

for them with courage and devotion. The forces we confront seem so great and time so short, it's easy to fear that our efforts are too scattered to be of real consequence. We tend to fall into the same short-term thinking that has entrapped our political economy.

The Great Turning invites us to lift our eyes from the cramped closet of short-term thinking and see the larger historical landscape. What a difference it makes to view our efforts as part of a vast enterprise, a tidal change commensurate to the crisis we face. What is underway, as many have observed, is a revolution that is comparable in magnitude to the agricultural revolution of the late Neolithic and the industrial revolution of the past two centuries. As the industrial-growth society spins out of control, the third revolution has arrived, which is even now given names like the ecological or sustainability revolution or the Great Turning. While the first two revolutions, as former EPA administrator William Ruckelshaus reflects, "were gradual, spontaneous, and largely unconscious, this [third] one will have to be a fully conscious operation. . . . If we actually do it, the undertaking will be absolutely unique in humanity's stay on Earth."

As compass, the Great Turning helps us see the direction in which our political economy is heading. Because the industrial-growth society is based on an impossible imperative—limitless increase in corporate profits—that direction leads to collapse. No system can endure that seeks to maximize a single variable. Already, our system is on "overshoot," using up resources beyond Earth's capacity to renew and dumping wastes beyond Earth's capacity to absorb. The losses inflicted on the biosphere now affect every system essential to life and deplete the diversity required for complex life forms to flourish. Yet, life is a dynamic process, self-organizing to adapt and evolve. Just as it turned scales to feathers, gills to lungs, seawater to blood, so now, too, immense evolutionary pressures are at work. They are driving this revolution of ours through innumerable molecular, intersecting alterations in the human capacity for conscious change.

Still, as Earth's record attests, extinctions are at least as plentiful as successful adaptations. We may not make it this time. Natural systems may unravel beyond repair before new sustainable forms and structures take hold. That is part of the anguish that is widely felt.

That anguish is unavoidable, if we want to stay honest and alert. The Great Turning comes with no guarantees. Its risk of failure is its reality. Insisting on belief in a positive outcome puts blinders on us and burdens the heart. We might manage to convince ourselves that everything will surely turn out all right, but would such happy assurance elicit our greatest courage and creativity?

The Great Turning, as a compass pointing to the possible, helps me live with radical uncertainty. It also causes me to believe that, whether we succeed or not, the risks we take on behalf of life will bring forth dimensions of human intelligence and solidarity beyond any we have known.

THE SCENE ON THE GROUND

This third revolution of our human journey is not only a possibility; it is a present, ongoing, multifarious phenomenon. The Great Turning is like a lens through which we can perceive the extent to which the revolution is happening. This lens is crucial, because it reveals developments that are ignored or distorted by the mainstream corporate-controlled media. In the words of Gil Scott-Heron, "The revolution will not be televised." It is hardly in the interests of billion-dollar industries, or the government that serves them, that we should know how they are being challenged and supplanted by grassroots initiatives.

These initiatives are sprouting on all sides, like green shoots through the rubble of a dysfunctional civilization. The Great Turning lens reveals that initiatives as different in character as a wind farm, a lawsuit against election fraud, and a fleet of kayaks protecting marine mammals are all part of a historic transition. It is important to review the three dimensions of this transition because they make it easier to see the Great Turning in action and to recognize our part in it. While presented as first, second, or third, they are not to be taken as sequential or ranked in importance. They coarise synergistically and are mutually reinforcing.

The first dimension includes all the efforts currently underway to slow down the destruction being wrought by the industrial-growth society. These range from petitioning for species' protection, to soup kitchens for homeless families, to civil disobedience against weapon makers, polluters, clear cutting, and other depredations. Often discouraging and even dangerous, work in this dimension buys time. Saving some lives, some ecosystems, some species, and some of the gene pool for future generations is a necessary part of the Great Turning. But even if every battle in this dimension were won, it would not be enough. A life-sustaining society requires new forms and structures.

The arising of these new forms constitutes the second dimension. Here we see the emergence of sustainable alternatives, from solar panels to farmers' markets, from land trusts to cohousing, perma-culture, and local currencies. At no other epoch in our history have so many ways of doing things appeared in so short a time.

Many of them—as in health, animal husbandry, and pest management—reclaim old, traditional practices. As promising as they are, these forms and structures cannot survive without deeply rooted values to nourish them. To proliferate and endure, they must mirror who we are and what we really want. They require, in other words, a profound change in our perception of reality.

This is the third dimension of the Great Turning: a shift in consciousness. Both personal and collective, both cognitive and spiritual, this shift comes through many avenues. It is ignited by the new sciences and inspired by ancient traditions. It also arises as grief for our world. Irreducible to private pathology, this grief gives the lie to old-paradigm notions of the isolated, competitive self. It reveals our mutual belonging in the web of life.

Now, in this very time, these three rivers—anguish for our world, scientific breakthroughs, and ancestral teachings—flow together. From the confluence of these rivers we drink and awaken to what we once knew: we are alive on a living Earth, source of all we are and know. Despite centuries of mechanistic conditioning, we want to name, once again, this world as holy.

Whether they come through Gaia theory, systems theory, chaos theory, or through liberation theology, shamanic practices, or the Goddess, such insights and experiences are absolutely necessary to free us from the grip of the industrial-growth society. They offer us nobler goals and deeper pleasures. They redefine our wealth and our worth, liberating us from compulsions to consume and control.

So rich is the harvest, that when we claim these new understandings, there's little room for panic or self-pity. Instead, gratitude arises to be alive at this moment, when, for all the darkness coming upon us, blessings abound. They help us stay alert and steady, so we can join hands to find the ways the world self-heals—and see the present chaos as seedbed for the future.

Among such blessings for me now, I count the explorations my colleagues and I are making into the mystery of time. With "deep time" practices, we enliven our felt connections with past and future generations and open our hurried fragmented lives into vaster expanses of time. The ancestors, who bequeathed us life, become more present to us, and so do the future ones, whom we carry within us like seeds.

These practices, long a feature of my workshops, gave rise to an extraordinary event last year. In Australia, where the Dreamtime is still a reality to the aboriginals who welcomed us, several dozen of us gathered to devote a full lunar cycle to immersion in deep time. The event was called Seeds for the Future: Training for the Great Turning.

There, under the wheeling stars by the southern sea, we felt the power of this planet-time. In our silence, rituals, and role playing, we sensed the ancestors and the future ones moving in our midst, encouraging us in the work that is ours to do. In our discussions, we felt the presence of those living now and the magnitude of their manifold efforts on behalf of life. Earth Community became for us not only a promise, but a present reality.

Returned to our daily lives, we call each other seedlings. That's what the Great Turning makes of us: seedlings of the future. How can I falter now, with so many hands and hearts at work, and all generations lending their support?

You Were Born for Such a Time as This

JAMES O'DEA

It has been said that 2012 presents a rare opportunity in the history of humanity to take inventory of what does not work—and to change it. James O'Dea, president of the Institute of Noetic Sciences, believes 2012 is our pay date, "our due date when our debt will be maxed and no further credit from a depleted and overtaxed Nature will be extended to us." In the following essay, he delves deeply into the dangerous state we've put ourselves and the world in, and heralds the time of healing to come. We have within us the potential for a great evolutionary change, he says, as he answers some of the essential questions of this new phase of development: What shape will this evolutionary shift take? How do we embrace our interconnectedness and cultivate the collective unconscious? How is it that we were born specifically for 2012 and this profound transition?

Life has rhythm and cyclical movement. Cycles of gestation, fruition, and dissolution constitute the very center of existence and the heart of our storied universe. Our greatest rituals, historical enactments, and mythic questing are pulsed by the mystery of time itself.

Time's mysteries are not, however, impenetrable. Those who have deciphered some of its majestic patterns and cosmic grandeur, from our wiser forbears to present-day elders and wisdom keepers, have understood that time becomes illusion if it is not wedded to the meaning and purpose of life. To delve into the mysteries of time, we must delve into the mysteries of life itself. And that, as they say, requires us to get a grip on reality.

To begin with, many of us have now come to appreciate that human life can only be understood within the context of all life, and that its origins were seeded in the birth of this universe. The limited idea that Earth was once the center of the universe has long been supplanted, but human beings still have considerable difficulty putting themselves in perspective in relation to the whole. While we can easily conjecture that nothing quite as entertaining as life on Earth appears to be going on in our neighborhood of this solar system, the billions of suns and myriad galaxies caught in the eye of the Hubble telescope have begun to shift our sense of cosmic scale. More and more people have begun to surmise that life does not have an exclusive engagement on Earth. In fact, no reasonable mathematical computation leaves us alone as the solitary expression of life in the universe.

Of equal or even greater significance is an ever-increasing sense that it is not helpful or accurate to think of a universe cluttered with dead matter and exploding gases randomly producing a perfectly ordered ecosphere for the exclusive rights and privileges of two-legged beings. The entire universe, we now understand, is finely tuned, and every millisecond of its birth, calibrated to allow for creation to unfold. Scientists fall to their knees in awe when they calculate the overwhelming precision of this universe. Impeccable timing seems to have been part of the design, and life as we know it is intricately woven into a cosmic movement that began billions of years ago. Life as we know it and time as we know it are inseparably linked.

2012 IS HUMANITY'S PAY DATE

Yet, for all intents and purposes, our world is not organized around this truth: time within the universal cycle is ignored, and society's clock obeyed—the clock that measures success by the Dow Jones Index, economic growth rates, and the gobble rate of mergers and acquisitions, which in several instances dwarfs the economies of nation states. Clock-time has been accelerating with such velocity that people's lives are increasingly shattered by stress, exhaustion, and the impossibility of fulfilling the multiple demands now made on them. We struggle to prevent our lives from fragmenting. We are increasingly spun-out. The more we are spun, the more our world seems to career out of balance, and our loss of rhythm commits us collectively to a race against time that is unsupportable and unsustainable. It is a race we cannot win: for time itself and the very purpose of life cannot be squeezed into the narrow funnel of the ego's feverish and furious enticements.

Just as we have not collectively organized our societal systems around the larger truths that shape and define human existence, we have not reached critical mass with regard to fully appreciating the looming consequences of the collective rampage created by the relentless pursuit of narrow self-interest. But we can take some consolation knowing that more people are waking up to the fact that, having devoured our own Earth habitat with unrestrained and often unconscious voracity, we are about to get the bill. The questions now squarely before us are: Will enough people wake in time? What will wake them? and How will they change?

This is one aspect of the meaning of 2012: it is humanity's pay date—the due date when our debt will be maxed, and no further credit from a depleted and overtaxed Nature will be extended to us. It is also the time when Nature's wake-up call, already sounding loudly, will reverberate though the psyche of humanity as a collective experience, as a shared frequency or resonance. I do not make this assertion based on shamanic divination, ancient prophecies, or the confluence of various calendars, as much as I note these with respect and interest. I make it because I live in these times.

I see the evidence of vanishing forests, shrinking water tables, melting ice caps, disappearing species, growing deserts, degraded oceans, toxic pollution, and signs of climate imbalance. I see belief systems in their death throes, distracted masses, delusional leaders, and incensed and humiliated peoples glorifying war and slaughter. I see the face of greed with its ever-expanding range of seductions, and I see the face of poverty with its multiple aspects of needless suffering and neglect.

I see all this without judgment—or without being judgmental, to be more precise. I see that I am a part of all this; it is my world. I am a part of the problem, and I am a part of the solution. It is because I choose not to live in denial, because I choose to be a witness to these times, with all their peril and promise, that I can see the implications of the swirling eddies of time itself as we move closer to 2012.

You don't have to proselytize me with how bad things are or how much worse they are going to get. I have been in the places where war, massacres, and state-sponsored violence have occurred. I am intimately acquainted with how cruel, intolerant, and exploitative human beings can be. I am equally acquainted with how governments rationalize their addiction to bigger and better weapons with all the trumpeted excuses of violent bigots and their legions of assassins. I have met the victims. I have met those who were beaten for their work in preventing rape and violence against women and the trafficking of children, bludgeoned because they sought to protect the rights of working people, or brutalized because they sought to end racism, torture, killing, and genocide. You don't have to persuade me that corporate elites control many of the levers of power in our world or that the media generally miss the story. You don't have to tell me because I choose not to live in denial. I am aware of the time in which I live, and for that reason, I see not only how bad things are but just how good they can become.

BEYOND THE THRESHOLD OF 2012
Hold this thought:

> As bad as it can get, which is extremely bad, the course of human evolution may also be close to a phase shift that will see the fulfillment of our deepest and most potent dreams. Right here, now, on the brink of calamity, a real possibility exists that we will discover that evolution has been gestating in us all the creativity and wisdom needed to transform our most dysfunctional beliefs. Ahead may lie the alignment of inner life and outward action and the emergence of a picture of reality shared by both science and spirituality.

Do not come to me with all the great insights into human capacities and the tantalizing insights of quantum physicists, brimming with affirmations about how we create our own reality, if you deny your own connection to the great wound in Nature and the great suffering of humanity. It is not someone else's nightmare; it is ours. What I want to know is, when looking unflinchingly into the painful truth of all that besets us in this hour of our collective history, can you also see why some of us may be looking to 2012 and beyond with a powerful sense of optimism? Can you hold *both* the meaning of the nightmare *and* the signs of our collective awakening—because the only way to get a grip on reality is to see that it is indivisible, reflected in *both* the shadow *and* the light, the bitter *and* the sweet. Time reveals the true nature of things, for time is the unalterable process that lets us see things for what they are and not simply what they appear to be in any given moment.

Can we appreciate what time it is on planet Earth? Beyond the terror, can we feel the awe, the power, and the beauty of what may lie beyond the threshold of 2012?

- Are you one of those people who wake with an irrepressible sense of purpose and meaning in your life, which no crisis, however bad, can dull with numbness or despair?
- Do you have a sense that you were born for such a time as this?
- Do you experience an embodied and enlived sense that a deeply humanized and peaceful planetary civilization will actually prevail on Earth?

If the answer is "yes, yes, yes," it is because all this has been gestating in you for as long as you can remember. You know that what is being born in your being is much older than you and that it has been a seed in the human imagination since ancient times; you are the dream of peace longing to be, longing to manifest its forms and find its fullest expression. Time has carried you to this fulcrum of becoming. The nightmare is not over, but you have begun to learn how to interpret the dramatic and portentous communication that it offers from the great mystery. As much as you feel that the husk of old forms is now destined to crack and break, you feel all the way to your gut that the seed of our higher nature is equally destined to break through.

Just how viable are you as a seed of future possibilities? Are these feelings and intuitions a symptom of naive or even cockeyed optimism, or are they a more accurate reflection of some deeper reality? Let's take a closer look.

To even conceive of planetary peace, one must imagine that people will be able to let go of—and yes, fully release—any enmity toward those who have perpetrated past crimes. Overcoming the fundamental sense of separation created by judgmental obsessions is a significant evolutionary signpost—one we should celebrate as profoundly liberating.

If you live in a place of sustained hopefulness about the human condition, it is probably because you yourself do not feed off antagonism for enemies. Something has shifted in you. You do not draw meaning or life force from polarized thinking. The shift in your awareness is, I suspect, one that has overcome inherited tendencies to hold on to judgmental feelings or fixations about others.

THE GREAT SHIFT FROM PUNISHMENT TO HEALING

When the most rudimentary moral categories of right and wrong are used as the rationale for violent aggression and societal oppression, this duality becomes the central organizing principle for polarizing forms of inclusion and exclusion. Causality is reduced to a series of proofs involving blame, and once culpability is determined, it is almost invariably the precursor to punishment. The good are rewarded and the bad punished without any deeper examination of the true complexity of the causal weave: in this worldview, perpetrators of bad deeds arise out of the blue and have no causal links to anything other than their own malevolence. This simple duality serves only to perpetuate the power of those who wish to maintain status-quo conditions. The more transformational stance is not guided by artificially circumscribed categories of who is right or who is wrong, but rather by looking at who among us is hurt or more deeply wounded and how they can find healing. There is no denying that wounds can form dangerous pathological behaviors—we see the evidence of that everywhere—but misidentifying wounds as sourced by an evil that must be punished or avenged does not make them go away.

Once the punitive and polarizing paradigm of right versus wrong is replaced with one of wounding and healing, judgmental posturing also fades away because we are no longer concerned with proving who is superior and who is inferior, but with restoring balance, reconciling differences, and restoring health. Can we see how our entire societal worldview and the fundamental constructs of international relations would be changed, or even irreversibly transformed, by an orientation guided by healing strategies and modalities? When is the day coming when one of the primary qualifications for leadership and public office will be a capacity to heal, to engage in deep reconciliatory dialogue that demonstrates genuine skill in moving beyond enmity to collaboration and the achievement of common objectives?

If 2012 represents the cresting of the breakdown, the death knell of empire, the graphic exposé of irrational fundamentalism, the inevitable ecological reckoning of systemic patterns of overconsumption, it also represents a decisive turning point when we have no other option than to seek healing and to collaborate our way out of chaos.

Between 2012 and 2020, in every arena, we will see the emergence of a great experiment in collective healing. Science and spirituality will form an unprecedented collaboration to address the root causes of pathologies that threaten to interrupt life on Earth—or that at the very least destine us to continued levels of trauma and suffering. It will be a collaboration that provides us with greater insight into the nature of the reality that pervades the cosmos. Across the planet, evolution will have coaxed us into a widespread recognition that our greatest capital is to be found in wisdom itself.

Now this is, of course, conjecture.

A QUICKENING IN OUR COLLECTIVE CONSCIOUSNESS

We do not know with certainty what will actually occur tomorrow, let alone what will happen across the planet in the years ahead. What we do know is how to:

- Read the contents of our own hearts and minds
- Discern in each other's eyes an undiminished capacity for compassion
- Heed our persistent capacity for truth-telling in the face of unprecedented manipulation
- Cultivate forgiveness for those who have harmed us or sought to suffocate our dreams

We can also notice that those who are more loving and conciliatory do not try to force their truth upon us and have learned to express their truth in ways that still allow for others' truth to be heard and felt. We can see in our daily experience that there are, in fact, many people who carry in their presence a magnanimity and life-enhancing energetic that invites tolerance, spaciousness for difference, and a capacity to be comfortable with ambiguity. When we look closely, we find a story unfolding within consciousness itself that is subtly taking shape and transforming our world from the inside out.

From this interior place, like a great root system growing invisibly beneath the surface, you and many others are having a potent sensory experience of much more love present in the world than is reflected in the dominant institutions or media. You are coming to know, in a primary, experiential way, that love is at the heart of evolution and the source of healing. Love is an evolutionary force because it is the primogenitor of nurturance. We live at a time when many people are discovering that the materialist paradigm, with all its wonderful attractions and distractions, is still unable to feed the more fundamental human longing for nurturance and relationship. Now, so-called hard-core pragmatists and realists are likely to pooh-pooh these notions. Familiar with their arguments, we don't have to rehash them.

> *The phenomenon we are discussing is not an ideological counter to the status quo; it is to be understood more accurately as a quickening in the center of our being—a quickening less susceptible to fear arousal, not conditioned by the gratifications of competitive advantage, but stimulated by experiences of wholeness, unity, and interconnection. It is what philosophers and theologians call the emergence of nondual consciousness.*

In recent decades a science has been emerging that supports the strength and power of this loving and generous inner reality. Forgiveness, it tells us, is healthy. Anger, resentment, and isolation constrict the heart and breed corrosive stress and depression. We live longer and healthier lives if we sustain deep relational ties to loved ones. Meditation, altruistic service, and gratitude are profound sources of wellness. Positive attitudes help us surmount loss and trauma. We thrive on affirmation, recognition, and respect. We blossom when we are appreciated and encouraged. We have learned that we have the capacity for extraordinary emotional intelligence and that our ability to tap the wellspring of creativity is greatly enhanced when we are nurtured and accepted and feel the presence of love in our hearts. Healing is still a mysterious process, but the evidence suggests that disease can sometimes be arrested, psychological affliction overcome. Even cycles of abuse can be transformed when we unleash the resilient powers of love and compassion and let go of fear, guilt, shame, regret, and the addiction to punitive vengeance.

Though I have seen mind-searing, heartbreaking levels of cruelty and indifference, I have also seen survivors of torture, oppression, and genocide demonstrate luminous capacities for reconciliation and transformation. I have seen perpetrators and victims forgive and reaffirm their common humanity. I have seen the indomitable nature of the human spirit, tested by all that pathological hatred can unleash, nevertheless rise above to find higher ground. This is the triumph of reality over delusion and the triumph of higher reason over narrow self-interest. It represents the way forward.

A GRAND CONJUNCTION UNFOLDS

This does not mean that we should delude ourselves into thinking that the way forward for humanity is guaranteed by a loving reality that seeks to remind us we are capable of living in alignment with its deeper truths. We know that those who dismantle international laws, hold prisoners incommunicado, torture, defend their economic interests at all costs, profit from military conflicts, supply lethal weapons to all comers, deny basic freedom, and, with insane disregard for human life, might use nuclear weapons and dirty bombs on large civilian populations—we know that they have and will resort to extreme measures, very extreme measures. We also know that social pathology can spread faster than wildfire with disastrous consequences. What is unique now is that the world is increasingly interdependent. We are in an irreversible global game. Whatever the scale of the breakdown—or breakthrough—that emerges from tensions and conflicts that could unravel over climatic imbalance, epidemics, resources, cultural domination, or religious fundamentalism, it will be a *global* breakdown—or a *global* breakthrough.

Thus, time and life on Earth are arriving at a grand conjunction. The frontiers of science must collaborate closely with spirituality to advance individual, social, and global transformation. The task of both science and spirituality will be to help us shed the more destructive fictions of our progress and at the same time illuminate the nature of reality. Evolution provides us with frameworks of meaning that are constantly in movement, even as the great underlying principles remain fixed. As we become clearer about those underlying principles, we learn to let go of what is more transitory and illusory. Our collective learning is fraught with difficulty and challenge, yet in the end, we are up to the challenge. Some psychologists and neuroscientists even say that more than intellectual achievement, what defines human progress is our capacity to meet and overcome challenges.

Despite an inappropriate use of power and all the other negatives that arise out of our stumbling and groping for truth and meaning, we humans have been evolving. Universal principles of law, democratic systems of government, technological advances, international commerce, education, and health-care systems reflect positive and liberating achievements despite their many inadequacies. Even a few hundred years ago, it would have been inconceivable that so much freedom, security, food, medicine, and housing could be provided for so many billions of people. Although there are catastrophic failures and alarming consequences for the ways in which we humans have developed, we have seen persistent creativity in both civil and political life and the broad extension of economic, social, and cultural rights. We have a more universal and shared sense of what it means to be a human being. Our collective conscious has been evolving.

More than ever, people are seeking precious wisdom that has been passed down from generation to generation—wisdom cultivated by tenacious spiritual traditions and diverse indigenous peoples. In aggregate, we are collectively more psychologically developed, more pluralistic, and less racist than our forbearers. Mind-body health is no longer a theory; it is a movement. We have come to see how our attitudes and beliefs are central to our own and others' well-being. We have begun to glimpse how consciousness itself is causal.

Evolution has been thrusting us outward, stressing us with enormous levels of complexity, but it has also forced us to go deeper inside ourselves to help us make sense of all that is happening at high speed around us and in the world at large. It impels us to find ways to anchor into a more spacious and secure inner reality, to find the peace and sustenance needed for life in the twenty-first century. Faced with the velocity and momentum of change, we learn to release the peripheral and grasp more firmly what is essential.

Many indigenous people think of this juncture in history—and for some, 2012 in particular—as a time of cleansing. Their innate resonance with cycles and timing tells them that Nature must now deal with the accumulated toxins we see everywhere. But for the indigenous, Nature is not something other than human life; it is not to be conceived as a separate force coming at the end of time to judge and punish us for our sinful addictions to unsustainable progress. We are a part of Nature, a deeply related part of the whole design, an inseparable part of all the connective tissues of existence itself. We are inside Nature, and like a great mother who sees that her children have fouled themselves, have failed to come home when called, are getting up to serious mischief, and are in great peril, she is going to remind us what time it is, where we are, and who we are. And who *she* is.

This then is the time of her call, the call of Nature resonating through the spirals of countless galaxies, inviting us to come into presence, to full attention and a deep remembrance of our ties to the whole, ties that can never be dissolved. As humans we have been given a reflective consciousness and a capacity to witness the flow of life beyond the perimeters of the self and its particular needs and preoccupations. This capacity is so much more than an intellectual capacity to reason, analyze problems, and predict consequences. It is a capacity to feel the resonance of the wider universe, a more inclusive world, and to experience a deep reverence for all life.

We now know how intimately the body feels and senses the world around it. Our hearts orchestrate a symphony of electromagnetic and biochemical resonance when they feel the presence of love. The circuitry of our brains fires new neural pathways when it encounters fresh meaning and new insight.

Our consciousness is most deeply perceptive and lucid when it is spiritually immersed and compassionately involved in the unfolding of the full range of experience—in other words, when it is paradoxically both detached and engaged. This detachment is not a separating or uncaring stance; as many of the traditions point out, it is a deep surrendering and a releasing of the struggle of the small will, so that it is not confined by the logic of narrow self-interest. In this kind of surrender, we experience a profound trust that, below the surface of the visible world, there is a vitality, an inspiring aliveness, and even a source of greater guidance. As detachment nurtures our sense of mysterious connection to a story so much larger and so much more multifaceted and dimensional than our little local identity, we are freed to engage in serving the world with a greater sense of being representatives of that great story, rather than missionary partisans for some narrow aspect of it. Our ability to develop an inclusive compassion is greater now than it has ever been. Of course, we have far to go, but we see global responses to natural catastrophes and diseases; we see a rise in philanthropic ventures and hundreds of thousands of organizations promoting rights, ecology, and poverty alleviation and as many celebrating arts and culture.

GLOBAL HEALING

Now imagine this responsiveness multiplied across religious and cultural divides—not stuck in punitive approaches to crime and violation, but rather applying healing and restorative modalities, and also restoring local economies and ecologies, building communities skilled in dialogic approaches to conflict, and cultivating appreciation for diversity.

For some, it will be hard to see that humans can and will evolve in this way; polarization has a way of fixating our attention and pulling us into its fragmentation and division. The sheer scale of corporate control and military budgets can paralyze us with feelings of powerlessness, and the persistent shallowness of political life can lead us to cynicism. But, like buds that have held tight in their own growth, humanity will blossom. Our blossoming will be the emergence of a consciousness no longer trapped in the old dualities, but reaching and spreading out beyond all narrow beliefs, identities, and affiliations. Our blossoming will come from inside the corporate world as well as outside, from inside and outside of politics, from transformed models of activism and altruism. Our collective emergence will be defined by feelings of oneness and communion with "all our relations." It will be accompanied by feelings of great expansiveness and generosity. We will want to open to others, to fully greet them with an appreciation for the mutuality of all suffering.

When we are ready to open up to one another, the first thing that becomes visible is the wound. Wounds need to be cleaned before they heal. Our wounds and the wounds we have inflicted on Nature are one. The wound in Nature is gaping; it is time to heal it. What a rich time of healing lies ahead. It has, of course, already begun. Visualize engaged healing activity in every sphere and across the planet from 2012 to 2020, and you will get a taste of things to come.

You were born for such a time as this!

Notes

CHOICE POINT 2012

1. Gallenkamp, Charles. *Maya: The Riddle and Rediscovery of a Lost Civilization.* New York: Penguin, 1999.

2. Coe, Michael D. *Breaking the Maya Code.* New York: Thames and Hudson, 1999.

3. Argüelles, José. *The Mayan Factor: Path Beyond Technology.* Santa Fe, NM: Bear & Co., 1987. 145.

4. India Daily. "Computer models predict magnetic pole reversal in Earth and sun can bring end to human civilization in 2012— evidence of extraterrestrial help in our survival." www.indiadaily. com/editorial/1753.asp.

5. *Science.* Vol. 168, 1969, Fig. 4.

6. *New York Times.* July 2004.

7. *ESSA Technical Report IER 46-IES 1.* July 1967.

8. *Science News.* Vol. 14, June 12, 1993.

9. *Science.* Vol. 260, June 11, 1993.

10. Royal Society of London, 243: 67–92, 1950.

11. *Nature.* February 26, 1998.

12. *Reviews of Modern Physics.* Vol. 29, 1957.

13. Sitchin, Zecharia. *The Lost Realms: Book IV of the Earth Chronicles.* New York: Avon Books, 1990.

THE ORIGINS OF THE 2012 REVELATION

1. Coe, Michael D. *Mexico.* London: Thames and Madison, 1962. 100.

2. Huxley, Aldous. *The Song of God: Bhagavad-Gita.* New York: New American Library, 1959.

1. More commonly known as the *b'ak'tun*. This essay uses the ancient Maya name, pik, instead of the modern term.

2. There is debate among Mayanists concerning when the close of the thirteenth pik (b'ak'tun) period will fall. The December 21 date used in this essay relies on the so-called GMT 584,283 calculation that coincides with the 260-day ritual calendar still used in the Guatemalan highlands, placing the period completion day of 4 Ahau on the Northern Hemisphere winter solstice of 2012. Other serious researchers prefer the 584,285 correlation and place the 2012 date on December 23.

3. Monument 6 at Tortuguero in Tabasco, Mexico.

4. Based on a review of the relevant literature and interviews with members of more than a dozen different Maya linguistic communities in Guatemala, Mexico, and Belize.

5. Yukatek Maya maintained a calendar based on periods of 20 tuns (7,200 days)—a relic of the full Long Count—well into the colonial period.

6. Personal communication, July 2006.

7. McFadden, Steven. *Profiles in Wisdom*. Santa Fe, NM: Bear & Co., 1991. 229.

8. www.quetzasha.com

9. www.quetzasha.com/entrevista.htm

10. http://homeofthemaya.tripod.com/index.html

11. http://homeofthemaya.tripod.com/mayan_beliefs.htm

12. *Utiw* also means "coyote" in K'iche'.

13. Personal communication with Elizabeth Araujo, October 2006.

14. The word is in the K'iche' language.

15. "Earth Prophecies: An Evening with Mayan Elder Don Alejandro." www.wisdomwritings.com. April 7, 2005.

16. http://tribes.tribe.net/mayawisdom/thread/a3fa6d61-25b1-4c9a-bc15-15dba9d37b30

17. Ibid.

18. http://edj.net/mc2012/bolon-yokte.html

19. Lorenzo Kinil of Chemax (1930). In Sullivan, Paul. "Contemporary Yucatec Maya Apocalyptic Prophecy: The Ethnographic and Historical Context," diss., John Hopkins University, 1984. 291–95.

20. Sullivan, Paul. *Unfinished Conversations: Mayas and Foreigners Between Two Wars.* New York: Alfred A. Knopf, 1989. 162

21. Lizama Quijano, Jesús J. "*Las señales del fin del mundo: Una aproximación a la tradición profética de los cruzo'ob.*" ("Signs of the End of the World: An Approach to the Prophetic Tradition of the Cruzo'ob") at www.uady.mx/sitios/mayas/investigaciones/historia/tradicion.htm. February 8, 2005. My translation from original Spanish.

22. Ibid.

23. July 2006.

24. Perera, Victor, and Bruce, Robert. *The Last Lords of Palenque: The Lacandon Mayas of the Mexican Rain Forest.* Los Angeles: University of California Press, 1985. 49. Hachakyum is the primary Lacandon deity.

25. In the Mam language, this may be derived from the Spanish term *mozo*, used to refer to hired day laborers, who were often not from an indigenous background, in some Mayan areas.

26. Personal conversation with a Mam chmanbaj in mid-May of 2002. (*Chmanbaj* is the nonpossessed form of the Mam word *chman*, which refers to both grandfather and grandchild.)

27. Tape-recorded conversation with Mr. González on December 27, 1996.

28. Montejo, Victor. *Maya Intellectual Renaissance: Identity, Representation, and Leadership.* Austin: University of Texas, 2005. 120–122.

29. Personal conversation, August 2006.

30. According to Edward Cleary, Protestant denominations make up about 25 percent of the total population in Guatemala. "Shopping Around: Questions About Latin American Conversions," 2004. www.providence.edu/las/Brookings.html. This figure may be much higher for the Maya alone.

31. In Todos Santos Cuchumatán, belief in a coming *fin del mundo* (end of the world) is commonplace among those belonging to the town's three principal fundamentalist churches.

GETTING TO 2012

1. www.unfpa.org/adolescents/facts.htm

2. Ibid.

3. International Energy Authority and Deutsche Bank.

4. Kunstler, James Howard. "Swan Dive." *Clusterfuck Nation.* October 20, 2006. http://jameshowardkunstler.typepad.com/ clusterfuck_nation/2006/10/swan_dive.html

5. Chivian, Eric. "Environment and Health: 7. Species Loss and Ecosystem Disruption—the Implications for Human Health." *Canadian Medical Association Journal,* January 2001.

6. McAlary, David. Voice of America News. Washington, November 2, 2006.

7. Hansen, Jim. *Proceedings of the National Academy of Sciences.* Vol. 103. 14288.

8. Greenpeace International. "The Global Retreat of Glaciers."

9. Whitehouse, Mark. "As Home Owners Face Strains, Market Bets on Loan Defaults, New Derivatives Link Fates of Investors and Borrowers in Vast 'Subprime' Sector." *Wall Street Journal,* October 30, 2006.

10. Buffet, Warren.

11. Soros, George. *The Collapse of Global Capitalism.*

THE MYSTERY OF A.D. 2012

1. For more information on cultural creatives, see www. culturalcreatives.org

2. Bailey, Alice. *The Seventh Ray: Revealer of the New Age.* New York: Lucis Publishing Company, 1995. This is a compilation of the writings of Alice A. Bailey and the Tibetan master Djwhal Khul.

3. For beginning of Aquarian Age from an esoteric studies perspective, see the 2006 article by Phillip Lindsay, "Zodiac and Ray Cycles in Esoteric Astrology: The Beginning of the Age of Aquarius." www.esotericastrologer.org/EA%20Essays/34%20Age%20of%20Aquarius%20ZodiacRayCycles.htm

4. C. W. Leadbeater wrote a classic book on chakras in 1927. For more on chakra petals, go to http://en.wikipedia.org/wiki/Petaljchakra.

5. Steiner, Rudolf.

6. Details about the galactic center are available from a NASA report in 2000. http://science.nasa.gov/headlines/y2000/ast29feb 1 m.htm

7. For more information on Tesla's scalar waves, see the work of Dr. Konstantin Meyl at www.meyl.eu

8. For an introduction to global scaling, go to www.globalscalingtheory.com

9. More information about biophoton research is available at www.lifescientists.de

10. A brief introduction to sonoluminescence is available at www.pat.llnl.gov/N Div/sonolum

11. For thoughts and emotions affecting genetic expression, see the work of Dr. Bruce Lipton at www.brucelipton.com/article/mind-over-genes-the-new-biology

12. For details about the immense potential energies of the vacuum, see Hal Puthoff's "The Energetic Vacuum: Implications for Energy Research." www.ldolphin.org/energetic.html

13. Dr. Randell Mills's work with hydrinos and his theory of classical quantum mechanics can be found at www.blacklightpower.com

14. The concepts of magnetoelectrism come from Dr. William A. Tiller and his books and articles. http://tillerfoundation.com/TheRealWorldModSci.pdf

15. Dr. Dean Radin's book *Entangled Minds* discussed this phenomena. See also http://deanradin.blogspot.com

16. See *The Findhorn Garden* or Dorothy Maclean's description of work with the devas at www.cygnusbooks.co.uk/features/myworkdevasdorothymaclean.htm

17. Kotlikoff, Laurence. "Is the United States Bankrupt?" *Federal Reserve Bank of St. Louis Review*, July/August 2006. 235–249. http://research.stlouisfed.org/publications/review/06/07/Kotlikoff. pdf

THE ALCHEMY OF TIME

1. Weidner, Jay, and Bridges, Vincent. *A Monument to the End of Time: Alchemy, Fulcanelli and the Great Cross*. Mount Gilead, NC: Aethyrea Books. 1999. *Mysteries of the Great Cross of Hendaye: Alchemy at the End of Time*. Rochester, VT: Destiny Books. 1999. www.jayweidner.com

2. For more information, see Weidner, Jay, and Rose, Sharron. "Tolkien at the End of Time: Alchemical Secrets of the *Lord of the Rings*." *New Dawn*, no. 82, January/February 2004. Rose, Sharron. "Reflections on the Cycle of the Great Yuga." www. sharronrose.com

3. For more information on the luminous energy body, see my films *Healing the Luminous Body* with Dr. Alberto Villoldo, *ArtMind* with Alex Grey, and *Yoga of Light* with Sharron Rose.

4. See "Alchemical Kubrick: 2001: The Great Work on Film." www. jayweidner.com

WILD LOVE SETS US FREE

1. Ray, Paul H., and Anderson, Sherry Ruth. *The Cultural Creatives: How 50 Million People Are Changing the World*. New York: Three Rivers Press, 2000.

2. Hicks, Esther, and Hicks, Jerry. *Ask and It Is Given: Learning to Manifest Your Desires*. Carlsbad, CA: Hay House, 2004.

3. Edwards, Gill. *Wild Love*. London: Piatkus Books, 2006.

4. Jung, Carl.

5. Williamson, Marianne. *Enchanted Love: The Mystical Power of Intimate Relationships*. New York: Touchstone, 2001.

6. Coehlo, Paulo. *The Zahir: A Novel of Obsession*. New York: HarperCollins, 2005.

AN AWAKENING WORLD

1. *Hadith.*

2. Hitchinson, Thomas, ed. *Wordsworth Poetical Works.* "Intimations of Immortality from Recollections of Early Childhood," 1.76. London: Oxford University Press, 1969.

3. Brother Lawrence (d. 1691) was a Carmelite lay brother, author of *The Practice of the Presence of God: The Best Rule of Holy Life.*

4. Suspended above the palace of Indra, the god who symbolizes the natural forces that protect and nurture life, is an enormous net. A brilliant jewel is attached to each of the knots of the net. Each jewel contains and reflects the image of all the other jewels in the net, which sparkles in the magnificence of its totality.

2012, GALACTIC ALIGNMENT, AND THE GREAT GODDESS

1. For an excellent introduction to the wisdom teachings of ancient Egypt, see Isha Schwaller de Lubicz, *The Opening of the Way.* Rochester, VT: Inner Traditions International, 1981.

2. Fulcanelli. *Les Mysteres des Cathedrales.* Suffolk, England: Neville Spearman, 1977. 58.

3. Paul LaViolette's book *Beyond the Big Bang* (Rochester, VT: Park Street Press, 1995) not only presents a compelling theory of continuous cosmic creation from a modern scientific perspective, but also places it in the context of the mythic and esoteric teachings of ancient Egypt.

4. See the documentary *Earth Under Fire* (Boulder, CO: Conscious Wave, 1998) for greater insight into the work of Paul LaViolette.

5. Evidence for this 13-sign alchemical zodiac was discovered in the Bibliotheque Nationale de Paris in a dossier of the mysterious Priory of Sion entitled *Le Serpent Rouge.* For the English translation of this document, go to www.connectotel.com/rennes/serpmwch.html

6. For greater insight into the Shakti/Shekhina current in the Hindu, Tibetan, and kabbalistic traditions, see Sharron Rose, *The Path of the Priestess* (Rochester, VT: Inner Traditions International, 2002).

7. From a conversation with Egyptian scholar and metaphysician John Nichols.

8. Naydler, Jeremy. *The Temple of the Cosmos: The Ancient Egyptian Experience of the Sacred.* Rochester, VT: Inner Traditions International, 1996. 45–46.

9. See "Fulcanelli's Tree of Life and the Mystery of the Cathedrals" from the *The Mysteries of the Great Cross of Hendaye,* by Jay Weidner and Vincent Bridges (Rochester, VT: Destiny Books, 2003) for a captivating look at the ankh in the context of the Western esoteric tradition.

10. *Emerald Tablet.*

11. Steinsaltz, Adin, *The Thirteen Petalled Rose.* USA: Basic Books, Inc., 1980. 118. This is an exquisite book on the Jewish mystical view of reality.

12. For greater understanding of the Lurian Kabbalah, see Drob, Sanford, *Symbols of the Kabbalah.* Northvale, NJ: Jason Aronson, 2000.

13. Upon the crown of her head, Isis wears a *uraeus* serpent symbol of the awakened Kundalini Shakti fully ascended to the pineal center. In this context, it is interesting to note that in ancient Egyptian, one of the names of the uraeus is "Iaret," which is derived from the verb meaning "to rise."

14. "Once this reunion (Ka and Ba) had taken place, the deceased became an akh—literally, an 'effective one,' able to live on in a new, nonphysical body." James Allen, curator at the Metropolitan Museum, New York City.

15. In many of the paintings and sculptures of ancient Egypt, the hieroglyph most associated with Isis is the throne, which is placed upon her head. This signifies that she is the seat and resting place for the divine. From a kabbalistic perspective, the divine throne or holy chariot is the vehicle through which the divine effulgence, or Shekhina, descends to the Earth. It is also the vehicle in which the seeker ascends to the ultimate reunion. In the words of Adin Steinsaltz, "For someone to comprehend the secret of the Chariot means that he is standing at the very focal point of the intersection of different worlds. At this intersection he is given knowledge of all existence and transformation, past, present, and future, and is aware of the Divine as prime cause and mover of all the forces acting from every direction."

16. *The Thirteen Petalled Rose.*

17. These teachings on the fully realized being have their parallel in the "Rainbow Body" teachings of the Dzogchen tradition of Tibet and the Incan teachings on the emergence in 2012 of a new species, which they describe as "Homo Luminous."

THE ADVENT OF THE POST-HUMAN GEO-NEURON

1. *The Book of the Jaguar Priest: A Translation of the Book Chilan Balam of Tizimin*, with commentary by Maud Worcester Makemson. New York: Henry Schuman, 1951.

Index

Note: Locators in italics indicate figures, tables, or illustrations.

dimethyltrytamine (DMT), 344–345
Divine, the, 89–90, 91. *see also* divine
    feminine; divine masculine; divine
    self; God; Source, the
divine feminine, 248, 301
divine masculine, 301
divine self, 57, 62
djed pillar, 323
Dmitriev, Alexey, 341
DNA, 78–79, 164–165, 213–214,
    393–395, 397. *see also* epigenomics
Dolphin children, 168
Double Slit Experiment, 12
Dreamspell, 97
Drosnin, Michael, *The Bible Code*,
    259–262
Druids, 301
dualist mind, 88–89
duality, 88–89, 248–249
duration, 147–148
Durga, 301
Dvapara Age, 198–199, 207. *see also*
    Bronze Age
Dzogchen tradition, 435n17

Eagle Clan, 282–283
Earth, 160–162, 315, 317, 318, 323,
    333
    consciousness of, 329
    current situation of, 74–77
    magnetism of, 1–16
    relationship to the galaxy as
        encoded in the Mayan calendar,
        73–74
Earth Community, 409
Earth/Moon system, 318
Earth mother, 344–345
"Earth Shift", ix
eclipses, 49
economy, the, 119–120, 138–139, 159,
    169–170
    disruption of, 138–139
    new economic models, 169
Edinger, Edward, 356
Edmonson, Munro, 332

Edwards, Gill, 243–257, 438
ego, 58, 65, 247–248, 250, 381
    excision of the, 88–89, 92
    humbling of, 58–59, 62
    overidentificaiton with, 57
egoism, 58
Egypt, 97, 210–211, 217, 222, 285,
    309–327, 435n15
Egyptians, sacred science of the,
    309–327
Ehecatl, 340
Eighth Underworld, 81, 85–86
Einstein, Albert, 147, 148, 167, 189,
    229
Electrolux, 183
electromagnetic activity, 167, 398–399
Elgin, Duane, 237
Elijah, 259–261
Eliyahu of Vilna, 260–261
El Morya, 300
Emerald Tablet, 296, 322
end-date, 48–49, 65, 101, 379
end-time, 218–220, 222
energy, 201–203, 205–206, 248, 249,
    252
energy changes, 398–399
energy-consciousness, 247
enlightenment, 26–29, 87–88, 90–91,
    227–242. *see also* awakening
enpowerment, *288*
Enron, 178
environment, 133–144, 159
Environment Agency (U.K.), 178
environmentally sustainable
    businesses, 181–183
epigenomics, 164–165. *see also* DNA
equinoxes, precession of the, 199–210,
    200, 211–212. *see also* precession
Eschaton, 364. *see also* singularity
essence, 319
Europe, 159
Evangelic Christianity, 264–265
Everett, Hugh, 12–13
evolution, 18–20, 22–24, *31*, 75–77,
    121, 122–123, 329, 367–387, 423.

Contributors

ARJUNA ARDAGH is the author of *The Translucent Revolution* and several other books. His newest book and documentary film, *Awakening into Oneness*, explores the phenomenon of Oneness Deeksha that has generated so much interest in recent years. Arjuna is the founder of The Living Essence Foundation in Nevada City, California. He speaks, appears regularly in the media, and leads seminars, longer retreats, and trainings throughout the United States and Europe.

JOSÉ ARGÜELLES, PH.D., is the New Age spiritual leader for the Planet Art Network and the Foundation for the Law of Time. He holds a Ph.D. in art history and aesthetics from the University of Chicago and has taught at numerous colleges, including Princeton University and the San Francisco Art Institute. Artist, poet, and historian, Dr. Argüelles initiated the Harmonic Convergence of August 1987. He is the author of *The Mayan Factor, Time and the Technosphere: The Law of Time in Human Affairs, Surfers of the Zuvuya*, and *Earth Ascending*. When asked how he would describe himself, Dr. Argüelles responded, "As a revealer of the context."

GREGG BRADEN is a *New York Times* bestselling author and is internationally renowned as a pioneer in bridging science and spirituality. Following successful careers as a senior computer systems designer for Martin Marietta Aerospace and the first technical operations manager for Cisco Systems, Braden's work, published in 14 languages and 23 countries, includes *The Isaiah Effect, The God Code,* and his newest bestseller, *The Divine Matrix.* For more than 20 years, Braden has searched high mountain villages, remote monasteries, and forgotten texts to uncover their timeless secrets. It is those secrets, he believes, that hold the key to meeting the greatest challenges of today. Beyond any reasonable doubt, his work shows us that we have the power to reverse disease, redefine aging, and even change reality itself, by embracing the focused power of belief and human emotion as the quantum language of change. Visit his website at www.greggbraden.com.

CARL JOHAN CALLEMAN, PH.D., holds a doctorate in physical biology and has served as an expert on cancer for the World Health Organization. He began his studies on the Mayan calendar in 1979 and now lectures throughout the world. He is author of *Solving the Greatest Mystery of Our Time: The Mayan Calendar* and *The Mayan Calendar and the Transformation of Consciousness.* He lives in Sweden. To learn more about him and his work, go to www.calleman.com and www.maya-portal.net.

GILL EDWARDS is a chartered clinical psychologist and author of *Living Magically, Stepping into the Magic, Pure Bliss, Wild Love,* and *Life Is a Gift.* She is trained in metaphysics, shamanism, energy psychology, and energy medicine, as well as psychodynamic and family/systemic therapy. Her passion is for exploring the practical links between the seen and unseen realms and how reality creation is linked with unconditional love. Gill has run Living Magically workshops in the United Kingdom and internationally since 1990. She lives in the heart of the English Lake District and adores walking in the mountains as much as she loves writing and teaching. Visit her website at www.livingmagically.co.uk.

JEAN HOUSTON, PH.D., scholar, philosopher, and researcher in human capacities, is one of the foremost visionary thinkers and doers of our time and is regarded as one of the principal founders of the Human Potential Movement. She is the author of several dozen books; her most recent is *Jump Time*. Considered to be one of the world's great teachers, she is the creator and principal teacher of two schools—a mystery school of philosophical, psychological, and spiritual studies, and a school for social artists. She has lectured in more than 100 countries and worked intensively in some 40 cultures. As an adviser to several United Nations agencies, she is training leadership in many developing countries through her programs to bring vision, innovation, and creative solutions to social issues at every level of society. To learn more about her, go to www.jeanhouston.org.

BARBARA MARX HUBBARD has been a leading voice for innovative change in the world for the past 40 years. As a visionary, author, speaker, social architect, and cofounder and president of the Foundation for Conscious Evolution, and a former nominee for the vice presidency of the United States, she posits that humanity now, as never before, is on the threshold of a quantum leap. Her books include *Conscious Evolution: Awakening the Power of Our Social Potential, Emergence: The Shift from Ego to Essence, The Evolutionary Journey*, and *The Hunger of Eve and Revelation*. Her DVD documentary series is called *Humanity Ascending: A New Way Through Together*. To learn more about her work, go to www.BarbaraMarxHubbard.com.

JANOSH is a Dutch graphic designer who, since 2003, has created computer-generated art incorporating principles of sacred geometry. He owns a popular gallery in central Amsterdam, Janosh Art and Design, and regularly tours the United States and Europe, exhibiting his work in diverse locations, from San Francisco's Supperclub to the United Nations. To learn more about his work, go to www.janosh-art.com.

JOHN MAJOR JENKINS is an independent researcher who has devoted himself to reconstructing ancient Mayan cosmology and philosophy. His books include *Tzolkin: Visionary Perspectives and Calendar Studies, Maya Sacred Science, Maya Cosmogenesis 2012*, and *Galactic Alignment*. He has taught at Esalen Institute and Naropa University, and his work has been featured on the Discovery Channel and the History Channel. To learn more about his work, go to www.alignment2012.com.

LAWRENCE E. JOSEPH is chair of the board of New Mexico–based Aerospace Consulting Corporation. He is the author of several books, including *Apocalypse 2012: A Scientific Investigation into Civilization's End*, and has written for a number of major newspapers and magazines, including the *New York Times*, Salon.com, *Family Circle, Audubon,* and *Discover.*

JOHN LAMB LASH, teacher, mythologist, and author of five books, has been called the heir and successor of Joseph Campbell. A specialist in the World Ages, precession, and comparative star lore, he has lectured in the United States and Europe and has appeared on numerous documentaries. His most recent book, *Not in His Image*, offers an intimate view of the mysteries of pre-Christian Europe and proposes a Gaia-oriented cosmology based on the Gnostic myth of Sophia, the goddess who morphed into the Earth. In *Quest for the Zodiac*, he compares the real-sky constellations known to the ancients with the popular astrological signs and presents a new paradigm for destiny. He is co-originator and principal author of www.metahistory.org, a teaching site sponsored by the Marion Institute.

ERVIN LASZLO, a native of Budapest, Hungary, is Docteur des-Lettres et Sciences Humaines of the Sorbonne (Paris) and the recipient of honorary Ph.D.s from universities in the United States, Canada, Finland, and Hungary. He is the founder and head of the General Evolution Research Group, founder and president of The Club of Budapest International, president of the Private University for Economics and Ethics (Vienna), and editor of *World Futures: The Journal of General Evolution* and the corresponding book series. He has published more than 400 scientific and public-interest papers and articles, and 80 books, translated into some 20 languages. He also edited a four-volume encyclopedia. Laszlo has been awarded a variety of distinctions, including the Franz Liszt Prize, the Goi Peace Prize of Japan, and the Mandir of Peace Prize of Assisi. He was nominated for the Nobel Peace Prize in 2004 and 2005. To learn more about his work, go to www.clubofbudapest.org.

MEG BLACKBURN LOSEY, PH.D., is the hostess of the *Dr. Meg Show: Conscious Talk for Greater Reality*, the author of *Pyramids of Light*, *The Children of Now*, and the *Online Messages*. She is a regular columnist in *Mystic Pop Magazine* and a contributor to many other publications. Dr. Losey has recently served as a consultant to *Good Morning America*. To learn more about her work, go to www.spiritlite.com.

JOANNA R. MACY, PH.D., is a scholar of Buddhism, general systems theory, and deep ecology. She is a leading voice in movements for peace, justice, and a safe environment. Interweaving her scholarship and four decades of activism, she has created both a groundbreaking theoretical framework for a new paradigm of personal and social change, and a powerful workshop methodology for its application. Her wide-ranging work addresses the psychological and spiritual issues of the nuclear age, the cultivation of ecological awareness, and the fruitful resonance between Buddhist thought and contemporary science. The many dimensions of this work are explored in her books *Despair and Personal Power in the Nuclear Age; Dharma and Development; Thinking Like a Mountain* (with John Seed, Pat Fleming, and Arne Naess); *Mutual Causality in Buddhism and General Systems Theory; World as Lover, World as Self; Rilke's Book of Hours; In Praise of Mortality* (with Anita Barrows); and *Coming Back to Life: Practices to Reconnect Our Lives, Our World* (with Molly Young Brown). Joanna has also written a memoir entitled *Widening Circles*. To learn more about her work, go to www.joannamacy.net.

KARL MARET, M.D., is president of the Dove Health Alliance, a nonprofit foundation based in Aptos, California, connected with an international network of physicians and scientists focused on energy medicine and other subtle energy approaches to healing. He specializes in complementary medical modalities, functional medicine, and energy medicine. As partner in an educational company called Heart-Mind Communications, he teaches seminars and is coauthoring a soon-to-be-published book entitled *Awakening the Dialogue of the Heart*. He is a poet, artist, author, and lifelong student of the Ancient Wisdom traditions, including esoteric astrology. To learn more about his work, go to www. heartmindcommunications.com.

CORINNE MCLAUGHLIN, coauthor of *Spiritual Politics* and *Builders of the Dawn*, is the executive director of The Center for Visionary Leadership based in Washington, D.C., and the San Francisco Bay Area. She coordinated a national task force for President Clinton's Council on Sustainable Development and has taught politics at American University. She is a fellow of The World Business Academy and the Findhorn Foundation. Portions of her essay are excerpted from her forthcoming book, *Creators of a New World*. To learn more about her work, go to www.visionarylead.org.

JAMES O'DEA, president of the Institute of Noetic Sciences, was the director of the Washington, D.C., office of Amnesty International for 10 years, where he represented Amnesty International to the State Department, the White House, the United States Congress, and the World Conference on Human Rights. Subsequently, he spent five years as executive director of Seva, a nonprofit organization dedicated to international health and development issues in Latin America, in Asia, and on American Indian reservations. He has participated with the World Wisdom Council (WWC) in partnership with Ervin Laszlo. To learn more about his work, go to www.noetic.org.

CHRISTINE PAGE, M.D., has over 30 years of experience in the caring professions, as both a medical doctor and a homeopath. Through her work, she is committed to finding ways to enhance a state of well-being through enhancement of the intuition, listening to the wisdom of the body, and working in cooperation with the more natural elements of this world. She is an international speaker, and her five books include *Spiritual Alchemy, Frontiers of Health, The Mirror of Existence, Beyond the Obvious*, and the *Mind/Body/Spirit Workbook*. To learn more about her work, go to www.christinepage.com.

JOHN L. PETERSEN's government and political experience includes stints at the National War College, the Institute for National Security Studies, the Office of the Secretary of Defense, and the

National Security Council staff at the White House. In 1989, Petersen founded The Arlington Institute (TAI), a nonprofit, future-oriented research institute. An award-winning writer, Petersen's first book, *The Road to 2015: Profiles of the Future*, was awarded Outstanding Academic Book and remained on the World Future Society's (WFS) best-seller list for more than a year. His latest book, *Out of the Blue: How to Anticipate Wild Cards and Big Future Surprises*, was also a WFS best seller. He has also written papers on the future of national security and the military, the future of energy, and the future of the media. Petersen is a past board member of the WFS, a network member of the Global Business Network, and a fellow of the World Academy of Art and Science. To learn more about his work, go to www.arlingtoninstitute.org.

DANIEL PINCHBECK has written features for the *New York Times Magazine, Esquire, Wired, Harper's Bazaar, Village Voice, Salon,* and many other publications. He was a 1999–2000 fellow of the National Arts Journalism Program at Columbia University. He has also been a columnist for the *Art Newspaper of London* and an editor at *Connoisseur Magazine*. He is the author of *Breaking Open the Head: A Psychedelic Journey into the Heart of Contemporary Shamanism* and *2012: The Return of Quetzalcoatl*. He is currently the editorial director of *Reality Sandwich* (www.realitysandwich. com) and a columnist for *Conscious Choice Magazine*.

SHARRON ROSE, M.A., E.D.D., is a filmmaker, teacher, writer, choreographer, and Fulbright senior research scholar in world mythology, religion, and the sacred arts of theater, music, and dance. She is the writer/director of the feature-length documentary *2012: The Odyssey*, author of the award-winning book *The Path of the Priestess: A Guidebook for Awakening the Divine Feminine*, creator of the DVD *Yoga of Light,* and producer of the *Sacred Mysteries* DVD collection. For the past 30 years, she has worked in the fields of education and the esoteric arts to investigate, integrate, and impart the knowledge and wisdom of ancient and traditional cultures throughout the world. This work has taken her across the planet and has been disseminated through master classes, lectures, workshops, live performance, and film and video. For more information, go to www.sharronrose.com or www.sacredmysteries.com.

PETER RUSSELL, M.A., D.C.S., is the author of 10 books and producer of two award-winning videos. His work integrates Eastern and Western understandings of the mind, exploring their relevance to the world today and to humanity's future. He has degrees in theoretical physics, experimental psychology, and computer science from the University of Cambridge, England. In India, he studied meditation and Eastern philosophy and, on his return, took up research into the psychophysiology of meditation at the University of Bristol. He was one of the first people to introduce human potential seminars into the corporate field, and for 20 years worked with major corporations on creativity, learning methods, stress management, and personal development. His principal interest is the inner challenges of the times we are passing through. His books include *The Global Brain, Waking Up in Time,* and the most recent, *From Science to God.*

ROBERT K. SITLER is an associate professor of modern languages and director of the Latin American Studies Program at Stetson University in DeLand, Florida. He received his doctoral degree from the University of Texas at Austin after completing a dissertation on Maya-related literature under the guidance of the late Dr. Linda Schele. His travels include more than 30 years of regular visits to the Maya world, among a wide variety of Maya ethnic groups. He has a particularly close relationship with the Mam Maya community of Todos Santos Cuchumatán in Guatemala. He has published a number of articles on Maya and Maya-related literature. His current research focuses on the cultural phenomenon surrounding the 2012 date on the Long Count Maya calendar. To learn more about Sitler and Maya perspectives on 2012, visit www.stetson.edu/~rsitler.

GEOFF STRAY, author of *Beyond 2012* and *The Mayan and Other Calendars*, has been studying the meaning of the year 2012 for over two decades and has compiled the information on his website, www.Diagnosis2012.co.uk. As a 2012 expert, he has given presentations around the world.

LLEWELLYN VAUGHAN-LEE, PH.D., is a Sufi teacher and author of the *Naqshbandiyya-Mujadidiyya* Sufi Order. Born in London in 1953, he has followed the Naqshbandi Sufi path since he was 19 years old. In 1991, he moved to Northern California and became the successor of Irina Tweedie, author of *Chasm of Fire* and *Daughter of Fire*. In recent years, the focus of his writing and teaching has been on spiritual responsibility in our present time of transition and the emerging global consciousness of oneness (see www.workingwithoneness.org). He has also specialized in the area of dreamwork, integrating the ancient Sufi approach to dreams with the insights of modern psychology. Llewellyn is the founder of The Golden Sufi Center (www.goldensufi.org). His most recent books are *Spiritual Power* and *Awakening the World: A Global Dimension to Spiritual Practice.*

JAY WEIDNER is a writer, filmmaker, and Hermetic scholar. He is the coauthor of *Mysteries of the Great Cross at Hendaye: Alchemy and the End of Time* and *A Monument to the End of Time: Alchemy, Fulcanelli, and the Great Cross*. His documentaries include *Artmind: The Healing Power of Sacred Art*, featuring Alex Grey; *Healing the Luminous Body: The Way of the Shaman*, featuring Dr. Alberto Villoldo; *Healing Sounds*, featuring Jonathan Goldman; *The Secrets of Alchemy*, which discusses the Cross at Hendaye; and many more. For more information on his work, go to www.sacredmysteries.com.

Credits

"The Mayan Factor" was recreated from an interview with José Argüelles. Reprinted by permission of Sounds True and José Argüelles.

"The Nine Underworlds" by Carl Johan Calleman is reprinted from *The Mayan Calendar and the Transformation of Consciousness*, by Carl Johan Calleman and José Argüelles. Reprinted by permission of Bear & Company.

Minor portions of "2012 and the Maya World" by Robert K. Sitler appeared previously in "The 2012 Phenomenon: New Age Appropriation of an Ancient Mayan Calendar." *Nova Religio* 9.3 (Feb. 2006).

"The Birthing of a New World" by Ervin Laszlo is excerpted from *The Chaos Point: The World at the Crossroads*. Reprinted by permission of Hampton Roads Publishing Company.

An earlier version of Jay Weidner's essay "The Alchemy of Time" first appeared in *New Dawn Magazine*, No. 92. September/October 2005.

"The Bible Code" is reprinted from Lawrence E. Joseph's *Apocalypse 2012: A Scientific Investigation into Civilization's End*. Reprinted by permission of Morgan Road Books.

"Jump Time Is Now" was adapted from the Introduction of Jean Houston's *Jump Time: Shaping Your Future in a World of Radical Change*. Reprinted by permission of Penguin Group (USA), Inc.

An earlier version of "A Vision for Humanity" by Barbara Marx Hubbard appeared on www.thevisionproject.org.

An earlier version of "The Great Turning as Compass and Lens" by Joanna R. Macy appeared in "5000 Years of Empire," the Summer 2006 *YES! Magazine*. www.yesmagazine.org.

Other Titles in the
Sounds True 2012 Collection

The Mystery of 2012
Anthology

On this audio supplement, eight luminaries delve deeper into the mystery of 2012.

Spoken-Word Audio / W1373D / 4 CDs / 5 hours
ISBN: 978-1-59179-722-7
UPC: 600835-137324 / U.S. $29.95

Unlocking the Secrets of 2012
John Major Jenkins

Did the ancient Mayans predict that the "end of time" would occur in 2012 . . . or is that a modern misunderstanding of their intended message?

Spoken-Word Audio / W1179D / 3 CDs / 3½ hours
ISBN: 978-1-59179-613-8
UPC: 600835-117920 / U.S. $24.95

Reality 2.0
Daniel Pinchbeck

A first-hand account of Pinchbeck's quest for a "system upgrade" for human society.

Spoken-Word Audio / W1286D / 4 CDs / 3¾ hours
ISBN: 978-1-59179-666-4
UPC: 600835-128629 / U.S. $29.95